New Aspects in Physiological Antitumor Substances

1st Oncology Workshop of the Society for Research of Organo- and Immunotherapy, Munich (FRG), and the K. E. Theurer Research Laboratories, Stuttgart (FRG), in Ludwigsburg (FRG), 26th August 1983

[Oncology Workshop (1st : 1983 : Ludwigsburg, Germany)]

New Aspects in Physiological Antitumor Substances

Experimental and Clinical Studies with Xenogenic Peptides and Proteins

Editors
G. Gillissen, Aachen; *K. E. Theurer*, Stuttgart

Scientific Organization
Th. Stiefel, Stuttgart

87 figures and 36 tables, 1985

RC271
C5
O52
1983

S. Karger · Basel · München · Paris · London · New York · Tokyo · Sydney

National Library of Medicine, Cataloging in Publication
Oncology Workshop (1st: 1983: Ludwigsburg, Germany) New aspects in physiological antitumor substances: experimental and clinical studies with xenogenic peptides and proteins / 1st Oncology Workshop of the Society for Research of Organo and Immunotherapy, Munich and K. E. Theurer Research Laboratories, Stuttgart, FRG, in Ludwigsburg, 26 August 1983; editors, G. Gillissen, K. E. Theurer. – Basel; New York: Karger, 1985.
ISBN 3-8055-4002-7
1. Antineoplastic Agents – therapeutic use – congresses
W3 ON50 1st 1983n
QZ 267 O58 1983n
2. Neoplasms – drug therapy – congresses
3. Neoplasms – immunology – congresses I. Gillissen, G. II. Theurer, Karl III. Gesellschaft zur Erforschung der Makromolekularen Organo- und Immunotherapie IV. Forschungslaboratorien Karl Theurer für Organo- und Immunotherapie V. Title

All rights reserved.
No part of this publication may be translated into other languages, reproduced or utilized in any form or by any means, electronic or mechanical, including photocopying, recording, microcopying, or by any information storage and retrieval system, without permission in writing from the publisher.

© Copyright 1985 by S. Karger AG, P.O. Box, CH-4009 Basel (Switzerland)
Printed in Germany by Mühlberger KG, D-8900 Augsburg
ISBN 3-8055-4002-7

Contents

Preface VII
Vorwort VIII
List of Participants IX

General Part

Theurer, K. E. (Ostfildern): Chrono-Biorhythmic Therapy with Non-Specific Methods and Medicaments with a Target-Specific Effect 1
Theurer, K. E. (Ostfildern): Circadian Biorhythm in Relation to Other Vegetative Regulations 8
Porcher, H. (Stuttgart): Therapeutic Integration of Xenogenic Proteins and Peptides in Modern Oncotherapy 17

Experimental Studies

Jachertz, D.; Jachertz, B. (Bern); *Mai, G.* (Frankfurt): Study of the Efficacy of Organic Extracts on a Cell-Free System of Hela Cells 24
Voelter, W.; Friedl, Ch.; Kalbacher, H. (Tübingen); *Munder, P. G.* (Freiburg); *Widmann, K. H.* (Freiburg): Synthesis of Thymopoietin 32–36 (TP 5) and its Effect on the Growth of Fibrosarcoma 33
Munder, P. G. (Freiburg); *Stiefel, Th.* (Stuttgart); *Widmann, K. H.* (Freiburg); *Theurer, K. E.* (Ostfildern): Antitumoral Action of Xenogenic Substances in Vivo and in Vitro 44
Letnansky, K. (Wien): The Inhibition of Tumor Proliferation by Specific Factors Separated from the Maternal Part of the Bovine Placenta 59
Maurer, H. R. (Berlin): Selective Effects of Sulfated Organ Lysates on the Clonal Growth of Normal Hematopoietic and Malignant Stem Cells in Vitro 70
Mayr, A.; Büttner, M.; Pawlas, S. (München): Studies of the Stimulation of Non-Specific Defense Mechanisms by NeyTumorin®-Sol 80
Ketelsen, U. P. (Freiburg): Pilot Study on the Influence of a Biological Response Modifier (NeyTumorin®) on the Plasma Membrane of Human Tumor Cells (Wish) in Vitro – in Comparison with a Chemocytostatic Agent (6-Mercaptopurine) 86

Pharmacological and Toxicological Studies

Stiefel, Th. (Stuttgart): Influence of NeyTumorin®-Sol and Subfractions on the Growth Behavior of Tumor Cells in Vitro 100

Gillissen, G. (Aachen): Biological Investigations with NeyTumorin®-Sol and NeyTumorin® E-Sol 106

Hadam, M. R. (Hannover): Flow Cytometry and Surface-Marker Phenotyping Using Monoclonal Antibodies: A Combined Approach to Precisely Define the State of the Immune System 120

Clinical Studies

Röhrer, H. (Breisach): Xenogenic Peptides and Proteins in Myelo- and Lymphoproliferative Disorders... 147

Douwes, F. R. (Bad Sooden-Allendorf): Immunmodulation: A New Therapeutical Method in Cancer Treatment? 155

Kisseler, B. (Böblingen); *Stiefel, Th.* (Ostfildern): Cytobiological-Cytostatic Combination Therapy. A New Approach in Medicamental Oncotherapy 170

Lange, O. F. (Bonn): Pilot Experience with NeyTumorin®-Sol for the Treatment of Generalized Metastasizing Carcinomas of the Mamma 194

Bohnacker, K. H.; Krause, F. (Löwenstein): Macromolecular Organ Extract (NeyTumorin®) in the Treatment of Non-Small-Cell Bronchial Carcinoma and Metastatic Lung Disease. Preliminary Report 202

Klippel, K. F. (Celle): Active Immunotherapy in Metastasizing Hypernephroma 210

Vetter, N.; Muhar, F. (Wien): AIDS – Clinical Picture and Therapy – Report on Two Patients 221

Preface

Specialization in natural science and medicine made the establishment of a special Study Group for Organo- and Immunotherapy in Oncology desirable. Such a group was founded on the occasion of the First Tumor Workshop in Ludwigsburg, FRG, on August 26th, 1983.

The multiplicity of lecture themes held in this workshop holds promise for the future. Although we are far from discovering the "miracle weapon" against cancer, the first results presented and discussed in this workshop give cause for hope. The organism disposes of regenerating forces for the purpose of preservation and recovery of health. These forces must be activated. Nothing is more essential than making use of the physiological, molecular factors of healthy and phylogenetically similar individuals, in order to support the body's regenerating forces for the purpose of a non-toxic treatment of tumor diseases. On the other hand, these factors can lessen the negative effects of invasive methods "steel, rays and chemotherapy" (i.e., surgery, tumor radiation and chemotherapy). Simultaneously, the tumor-patient's subjective state can be permanently improved. Experimental research lends a solid foundation to furthering these results in hospital and medical practice.

Aachen and
Ostfildern, 1984

G. Gillissen
K. E. Theurer

Vorwort

Die Spezialisierung in der Naturwissenschaft und der Medizin ließ einen speziellen Arbeitskreis für Organo- und Immunotherapie in der Onkologie wünschenswert erscheinen. Dieser wurde anläßlich des Tumor-Workshops in Ludwigsburg am 26. 8. 1983 gegründet.

Die Vielfältigkeit der Vortragsthemen dieses Workshops weist in die Zukunft. Obwohl man noch weit davon entfernt ist, die Wunderwaffe gegen Krebs entdeckt zu haben, berechtigen die ersten Ergebnisse, die bei diesem Workshop vorgetragen und diskutiert wurden, zu begründeten Hoffnungen. Der Organismus verfügt über regenerative Kräfte zur Erhaltung und Wiederherstellung der Gesundheit; diese gilt es zu aktivieren. Nichts liegt näher, als die physiologischen molekularen Faktoren aus gesunden, phylogenetisch ähnlichen Individuen zur Unterstützung der körpereigenen Wiederherstellungsbestrebungen im Rahmen einer nichttoxischen Krebs-Therapie zu nutzen. Andererseits können solche Faktoren die nachteiligen Wirkungen der invasiven Methoden durch «Stahl-, Strahl- und Chemotherapie» vermindern und das subjektive Befinden der Tumorpatienten nachhaltig verbessern. Die experimentelle Forschung fundiert die vorliegenden Ergebnisse aus Klinik und Praxis.

Aachen und
Ostfildern, 1984

G. Gillissen
K. E. Theurer

List of Participants

Bohnacker, K., Dr. med., Klinik Löwenstein, D-7101 Löwenstein

Büttner, M., Dr. med., Institut für Med. Mikrobiologie, Infektions- und Seuchenmedizin der Ludwig-Maximilian-Universität München, Veterinärstr. 13, D-8000 München 22

Douwes, F. A., Prof. Dr. med., Ärztlicher Direktor der Sonnenberg-Klinik, Hardtstr. 13, D-3437 Bad Sooden-Allendorf

Gillissen, G., Prof. Dr. med. Dr. rer. nat., Institut für Mikrobiologie der RWTH Aachen, Pauwelsstraße 27–29, D-5100 Aachen

Hadam, M., Dr. med., Onkologisches Labor – Kinderchirurgie, Med. Hochschule Hannover, Karl-Weichert-Allee 9, D-3000 Hannover 61

Jachertz, D., Prof. Dr. med., Direktor des Instituts für Hygiene und Mikrobiologie der Universität Bern, Friedbühlstr. 51, CH-3010 Bern

Ketelsen, U. P., Prof. Dr. med., Universitätskinderklinik, Abt. Pädiatrische Muskelerkrankungen, Mathildenstr. 1, D-7800 Freiburg i. Br.

Kisseler, B., PD, Dr. med., Chefarzt der Radiologischen Zentralabteilung des Kreiskrankenhauses Böblingen, Bunsenstr. 120, D-7030 Böblingen

Klippel, K. F., Prof. Dr. med., Chefarzt der Urologischen Klinik, Allgemeines Krankenhaus, Siemensplatz 4, D-3100 Celle

Krause, F., Prof. Dr. med., Chefarzt der Klinik Löwenstein, D-7101 Löwenstein

Lange, O. F., Dr. med., Oberarzt der Robert-Janker Klinik, Baumschulallee 12, D-5300 Bonn

Letnansky, K., Prof. Dr. rer. nat., Institut für Krebsforschung der Universität Wien, Borschkegasse 8 a, A-1090 Wien

Maurer, H. R., Prof. Dr. rer. nat., Freie Universität Berlin, Fachbereich Pharmazie (FB 22), Königin-Luise-Str. 2–4, D-1000 Berlin 33

Mayr, A., Prof. Dr. med. vet., Dr. h. c. mult., Direktor des Instituts für Medizinische Mikrobiologie, Infektions- und Seuchenmedizin, Veterinärstr. 13, D-8000 München 22

Muhar, F., Prof. Dr. med., Chefarzt der II. Internen Lungenabteilung des Pulmologischen Zentrums der Stadt Wien, Sanatoriumstr. 2, A-1145 Wien

List of Participants

Munder, P. G., Prof. Dr. med., Max-Planck-Institut für Immunbiologie, Stübeweg 51, D-7800 Freiburg-Zähringen

Porcher, H., Dr., Kernerstr. 26, D-7000 Stuttgart 1

Röhrer, H., Dr. med., Oberarzt der Inneren Abteilung des Rosmann-Krankenhauses Breisach, Augustinerberg 8, D-7814 Breisach

Stiefel, Th., Dr. rer. nat., Olgastr. 139/2, D-7000 Stuttgart 1

Theurer, K. E., Prof. Dr. med., Brunnwiesenstr. 23, D-7302 Ostfildern 1

Vetter, N., Dr. med., II. Interne Lungenabteilung des Pulmologischen Zentrums der Stadt Wien, Sanatoriumstr. 2, A-1145 Wien

Voelter, W., Prof. Dr. rer. nat., Leiter der Abteilung für Physikalische Biochemie, Physiolog.-chem. Institut der Universität Tübingen, Hoppe-Seyler-Str. 1, D-7400 Tübingen

General Part

New Aspects in Physiological Antitumor Substances, pp. 1–7 (Karger, Basel 1985)

Chrono-Biorhythmic Therapy with Non-Specific Methods and Medicaments with a Target-Specific Effect*

K. E. Theurer

Research Laboratories of Organo- and Immunotherapy,
Ostfildern-Ruit, FRG

Knowledge of vegetative metabolism regulations of the stress mechanism, its circadian biorhythmic and system changes in metabolism using minimal stimuli are a fundamental part of any therapy with target-specific medications, as well as with non-specific methods [6].

The Significance of Blood Chemistry for the Formation and Decomposition of Tumors

According to *Leupold* [3], formerly Professor in Ordinary for Pathology at the University of Cologne, FRG, the forming of a tumor can depend on the metabolism of the tissue and on the total metabolism. Through the parenteral application of very small quantities of protein breakdown products of physiological origin, lipids, carbohydrates, and ions it was possible to inhibit or promote the proliferation of inflammatory new cell formations up to malignantly developing tumors. The cell environment of the organism, the liquids in contact with them, contains the most diverse regulatory factors. Similarly, the vegetative nervous system and the basic vegetative formation of the active mesenchyma [4], pervading everything, also have an influence on cell proliferation.

---

* Medsche Klin. *4:* 79 (1984).

This "cell environment system", despite a large measure of homeostasis, is subject to certain changes. It is said that the reaction tendency is mainly dependent on the course in time and the extent of movements of the system. When the quotient was lowered, inflammatory chronic proliferations, adenomas, sarcomas or carcinomas – especially of the kidneys, lungs, intestine, mamma, uterus, etc. – were observed in animal experimentation within a time-span in some cases of more than two years.

With system movements in the opposite direction to that which leads to tumors, i.e., when there is a rise in the quotient through an increase in the cholesterol and phospholipids and a decline in sugar, tumors are decomposed by necrosis. Benign tumors of the mamma, prostatic hypertrophy and cell proliferations with psoriasis, fibroses, infiltrates and fistulae also responded to such a quotient change.

Leupold performed the system movements by subcutaneous injection of physiological mixtures in a quantity of 0.05 ml. Larger injections displayed a lesser or no effect. With animals, the injection was more effective in the cerumen than in the back (possible relation to acupuncture of the ear). The treatment was carried out for a fairly long time at different intervals. In addition to the injection, mixtures of substances, especially a lipid solution, were administered perorally. The trauma of intercurrent surgery and radiation had the same effect on the blood chemistry in some cases as in animal experimentation, leading to the appearance of tumors and recidivation.

Sympathetic Tone Promotes the Forming of Tumors

The *Leupold* therapeutic methods consist essentially on one hand of the substitution of metabolic factors, whose highest concentration is achieved in a sympathicotonic reaction situation (glucose), and on the other, in a stimulus which triggers the desired system movement. For the decomposition of tumors, this corresponds to the parasympathicotonic phase and for the forming of tumors, to the sympathicotonic reaction phase. However, *Leupold* has not yet been able to account for these relations.

The system movement recommended for the decomposition of tumors can be enhanced by all the methods and medicaments with a parasympathomimetic effect. Favorable therapeutic conditions can also be created by the prior raising of factors at the end of the sympathicotonic reaction phase, which then decline during the parasympathetic preponderance.

In a research project comissioned in 1957, the Study Group for Cancer Control of North Rhine/Westphalia, FRG, carried out a study of the experimental fundamentals of *Leupold*'s theory [1]. The results obtained were not convincing, however. The curves of the metabolic parameters, which were said to lead to the formation of tumors, could indeed be reproduced but not their effect on the experimental formation of tumors. The system movement was the expression of a non-specific stress reaction, as can also be induced solely by the taking of blood samples.

The findings concerning the therapeutic consequences for human medicine were contradictory. Nevertheless, various hospitals described "reactions on tumors and patients to which there was no objective objection" and which "should give occasion to further study of the problem of the decomposition of tumors and the influencing of the metabolism" or "should encourage more intensive work on this approach" [1].

It is not our intention here to examine methodic differences in a follow-up examination as compared with the method used by *Leupold*. It is possible that the metabolic stimulus was not applied at the start of the change in the parasympathicotonic vegetative regulation. Also that no animals were used having a high incidence of tumors and with a syngeneic or genetic disposition for the formation of tumors. According to the "genetic cancer theory" [7] favored at the present time, it is probable that stress situations with a genetic cancer disposition favor the formation of carcinomas by reason of reduced adaptability as a result of the lability of the genetic regulations. This lability could then also have an effect on the sensitivity of tumor cells to therapeutic system movements.

Therapeutic Consequences: Circadian Parasympathetic Stimulation

Although the experimental formation of tumors as described by *Leupold* could not be confirmed and is possibly dependent, to a major extent, on genetic factors, the stimulation of the parasympathetic reaction phase appears to be a logical measure for patients with a malignant growth, but also for disorders of an inflammatory-infectious nature with reduced immune resistance – which is increased with parasympathetic preponderance – and also for all disorders with a mainly sympathicotonic reaction tendency.

A metabolic therapy matched to the circadian rhythm is also justified for other indications without a tissue proliferation tendency but with a parasympathicotonic reaction situation. These include both degenerative disorders with depositions, especially of calcium, cholesterol, lipids, uric acid, pigments and other metabolites, and also neurological and psychiatric conditions, especially Parkinson's disease, cramp and migraine, depressions and thought-blocking, and also allergic and immunopathogenic autoaggressive conditions of the rheumatic type (K-types according to *Curry* or A-types according to *Lambert* [5]). The therapeutic stimulus in this case should come at the start of the sympathicotonic reaction situation, i.e., between 3 a.m. and 7 a.m. or between 3 p.m. and 6 p.m.

Circadian vegetative reactions can thus be both agonistically and antagonistically, synchronously and also asynchronously influenced. A "shaking-up" of the vegetative rhythms also appears to be a promising approach for the withdrawal-treatment of addicts.

Determination of an Individual Vegetative Reaction Curve

For a treatment in keeping with the rhythm, it is necessary for the individual circadian rhythm to be identified. The metabolic tests recommended by *Leupold* are too involved for this. It is usually sufficient to record the changes per unit time in functional capacity; well-being, response, body temperature, appetite, sleep requirement and activity in the course of the day to be able to draw a simplified conclusion about a sympathicotonic and parasympathicotonic reaction situation. Simple methods of measurement of the "actual vegetative reaction situation" for the determination of a "vegetative reaction curve" [5], can also take the form of resistance measurements of the skin (*Hauswirth and Kracmar*'s R-C measurements [2]).

For the stimulation of the sympathicotonic reaction situation, use can be made of all those factors which are usually increased during this phase, and for the inhibition of all the factors which are higher than normal when there is a parasympathetic preponderance. Timed stimuli through light, music, application of heat and food intake at the correct time can synchronize a disturbed circadian rhythm with the environment, or enhance or inhibit individual phases.

Chrono-Biorhythmic Therapy

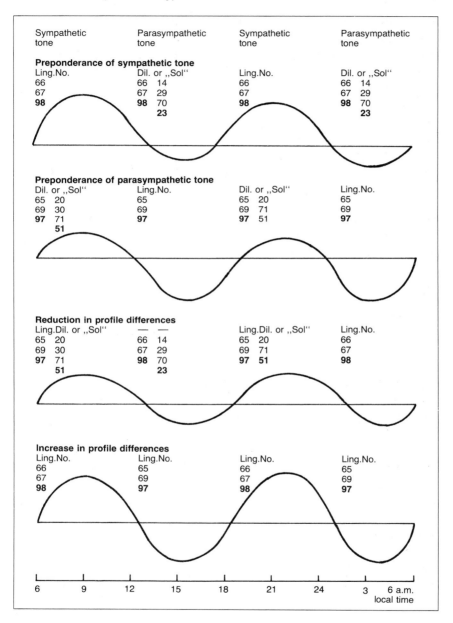

Fig. 1. Relations of certain organs to the parasympathicotonic or sympathicotonic phase.

Relations of Certain Organs to the Parasympathicotonic or Sympathicotonic Phase

From the comparison of the circadian fluctuations of physiological parameters [6], it is possible to allocate certain organs to the possible antagonistic reactions. A reduced sympathicotonic reaction situation is found with an insufficiency of the hypophysis, adrenal gland, thyroid gland, oestrogen hormone production, parathyroid gland; a reduction of the parasympathetic tendencies with an insufficiency of the androgenic and progestogenic hormone production and of the epiphysis, pancreas and thymus. When immunological principles are observed, suitable organ preparations can be applied repeatedly in keeping with the rhythm, as individual preparations, and in organ combinations. Special medications are available for this (fig. 1).

A direct influence by sympathicomimetics or parasympathicomimetics in the form of medicaments or hormones, like every monotherapy with a blocking effect, disturbs the regulations of the natural biorhythms when they encroach on the antagonistic reaction and can lead to disadvantageous side effects [6]. In contrast to this, cytoplastic extracts from organs [7] which favor a special reaction situation have an alterating and normalizing effect.

Summary

Every specific therapy has a non-specific active component. In order to attain optimal treatment results it is, therefore, necessary to take the non-specific mechanisms into consideration. A synoptic consideration of the principles of the non-specific therapy, in particular of the circadian rhythm, stress-adaption syndrome and total vegetative change-over with respect to blood-chemistry changes in the cell and tissue-metabolic processes, renders important a temporal and rhythm-adapted application of biological methods and medication. The change from a sympathicotonic, catabolic reaction situation to a parasympathicotonic, anabolic reaction situation, leads to changes in different parameters. Non-specific irritations (stress) can increase the intensity of particular courses of illness and lengthen their duration. Additionally, the system motions, such as of blood sugar or of O_2 partial pressure, can be intensified by rhythm-adapted substitution of glucose and insuline-oxygen respiration, and respiratory exercises. *Leupold* healed dermatoses, in particular the proliferous and chronically inflammatory type, through to benign and malignant growths, whenever he implemented a non-specific metabolic stimulant, as in the case of intensive sinking of blood sugar over a period of 2–3 hours and, if possible, while blood lipids were on the increase.

Reversing the process, he was also able, by using animal experiments, to produce corresponding diseases, presumably in the case of predisposed individuals. Through the observation of the circadian reaction changes, which were then not known to *Leupold*, the specific treatment methods, including organo- and immunotherapy, should be able to further increase the changes of succes.

Zusammenfassung

Jede spezifische Therapie hat eine unspezifische Wirkkomponente. Für optimale Behandlungsergebnisse ist es deshalb erforderlich, auch die unspezifischen Mechanismen zu berücksichtigen. Eine synoptische Betrachtung der Grundlagen der unspezifischen Therapie, insbesondere des Zirkadian-Rhythmus, des Streß-Adaptions-Syndroms und der vegetativen Gesamtumschaltung in Beziehung zum Blutchemismus bzw. den Änderungen im Zell- und Gewebsstoffwechsel, läßt eine zeit- und rhythmusgerechte Applikation auch biologischer Methoden und Arzneimittel wichtig erscheinen. Der Wechsel von sympathikotoner, kataboler Reaktionslage in die parasympathikotone, anabole Reaktionslage führt zu Veränderungen verschiedener Parameter. Unspezifische Reize (Streß) können die Intensität besonderer Abläufe verstärken und ihre Dauer verlängern. Zusätzlich können die Systembewegungen, z. B. des Blutzuckers oder des O_2-Partialdrucks durch rhythmusgerechte Substitution von Glukose und Insulin bzw. Sauerstoffbeatmung oder Atemgymnastik verstärkt werden. *Leupold* hat Heilungen von Dermatosen, insbesondere von proliferativen und chronisch-entzündlichen Vorgängen, bis hin zu gutartigen und bösartigen Geschwülsten erzielt, wenn er einen unspezifischen Stoffwechselreiz bei intensivem Absinken des Blutzuckers über 2–3 Stunden und möglichst im Anstieg der Blutlipide gesetzt hat. Andererseits konnte er tierexperimentell bei umgekehrtem Systemverlauf entsprechende Krankheiten, vermutlich bei prädisponierten Individuen, auslösen. Durch Beobachtung der zirkadianen Reaktionsveränderungen, die *Leupold* noch nicht bekannt waren, dürften sich auch bei spezifischen Behandlungsmethoden, einschließlich der Organo- und Immunotherapie, die Erfolgschancen weiter verbessern lassen.

References

1 Arbeitsgemeinschaft für Krebsbekämpfung Nordrhein-Westfalen: Überprüfung der wissenschaftl. Grundlagen der Lehre von E. Leupold: Kampf dem Krebs, Heft 3, 1962.
2 Hauswirth, O.; Kracmar, F.: Arch. phys. Ther. *11:* 6 (1959).
3 Leupold, E.: Der Zell- und Gewebestoffwechsel als innere Krankheitsbedingungen (a); Die Bedeutung des Blutchemismus besonders in Beziehung zu Tumorbildung und Tumorabbau (b) (Thieme, Stuttgart, a 1945, b 1954).
4 Pischinger, A.: Krebsarzt *21:* 297 (1968).
5 Rilling, S.: Erfahr. Heilk. *17:* 352 (1960).
6 Theurer, K. E.: Medsche Klin. *79:* 3 (1984).
7 Theurer, K. E.: Medsche Klin. *60:* 1909 (1965).
8 Theurer, K. E.: Ärztl. Prax. *23:* 1709 (1981).
9 Theurer, K. E.: Therapiewoche *33:* 17 (1983).

Prof. Dr. Karl E. Theurer, Forschungslaboratorien für Organo- und Immunotherapie, Brunnwiesenstraße 21, D-7302 Ostfildern-Ruit (FRG)

Circadian Biorhythm in Relation to Other Vegetative Regulations*

K. E. Theurer

Research Laboratories of Organo- and Immunotherapy,
Ostfildern-Ruit, FRG

It still holds true that little or no use is made in hospitals and medical practices of the many new discoveries in medical science, as they call for a turnabout in thinking or because the application of these findings appears complicated and inconvenient. The synopsis of related mechanisms should help to overcome any reservations there may be as to their integration in medical thinking.

Every specific therapy contains non-specific active components (table I). Non-specific effects on the organism as a whole can trigger, as well as heal, diseases. Up till now, this dualism had been attributed to different intensities of stress. In addition to the dosage, the site of application is also important. Another factor which is given little attention is that of the time of day. Even ancient Chinese medicinal applications knew of this dependence as mentioned in the Yin-Yang doctrine and the "organ clock" for the most effective timing for the functioning of individual organs [1].

Daily Periodic Course of the Reaction-Situation, the Metabolism and Hormonal Regulation

Research results on the chrono-biorhythm show a circadian change, i.e., a change in accordance with the time of day, in the homeostasis of the most diverse parameters in the organism [5]. To a large extent, these

* Medsche Klin. *3:* 79 (1984).

Table I. Synopsis of the fundamentals of vegetative regulations

1. Circadian rhythm of biological systems
2. Total vegetative alteration
3. Stress and adaptation syndrome
4. Cybernetics
5. Non-specific change-over by stimuli (physical, chemical, medicamentous) on the skin, muscles and joints, active mesenchyma (physical therapy, chirotherapy, massage, etc.), blood (ozone therapy, O_2-multiple-step therapy), nerves (neural and segmental therapy, acupuncture), metabolism (diet, weight reduction, physical movement, sport, climate)
6. Focal theory
7. Mental conditioning, autogenous training
8. Simultaneous non-specific effect of medicaments with a specific action
9. System changes of metabolism and hormonal regulation through minimal stimuli with metabolic mixtures and/or electrolyte solutions

changes are triggered by a change in the vegetative regulation between a sympathicotonic and a parasympathicotonic reaction. In addition, certain deviations also exist which can be caused by external factors such as nutrition, stress, etc.

The humoral changes in the various factors with a sympathicotonic and a parasympathicotonic reaction situation are set out in table II; no claim as to completeness. Intracellular changes are mostly contrary to this. Humoral changes in the mineral balance can be found from the quotient for neuromuscular excitability according to *György* [2]:

$$\frac{K \times \text{phosphates} \times HCO_3}{Ca \times Mg \times H}$$

In this formula, the numerator contains factors which rise with a preponderance of the parasympathetic nervous system and the denominator factors arising when a sympathetic nervous system predominates.

Any change in the individual circadian rhythm (desynchronization), e.g., through transcontinental flights across time-zones or asynchronous stimuli, medication or activity (nightwork), is experienced as unpleasant and can lead to signs of illness. Similarly, an intensification of the reaction changes by substitution of certain factors, a prolongation or shortening of individual phases can either cause illness or induce a therapeutic stimulus [4].

Table II. Comparison of circadian fluctuations of physiological parameters

Parameter	Reaction situation	
	sympathetic	parasympathetic
Anabolism	−	+
Catabolism	+	−
Blood sugar	+	−
− calcium	+	−
− potassium	−	+
− magnesium		−
− sodium	+	−
− phosphates	−	+
− lipids	−	+
− proteins	+	−
− acidosis	+	−
− alkalosis	−	+
− O$_2$-partial pressure	+	−
Myeloid tendency	+	−
Lymphatic tendency	−	+
− eosinophilia	−	+
− immunoglobulins	−	+
BSR	+	−
Body temperature	+	−
Elec. conductivity of the skin	+	−
BMR	+	−
Blood pressure	+	−
Pulse-rate	+	−
Hormones of		
− hypophysis	+	−
− adrenal	+	−
− thyroid	+	−
− oestrogens	+	−
− androgens	−	+
− parathyroid	+	−
− epiphysis	−	+
− pancreas	−	+
− thymus	−	+
− placenta decidua	−	+
− chorion	+	−

The Course of the Phase of a Complete Vegetative Change-Over

The circadian rhythm varies from one individual to another. There are early risers (larks) and night-birds (owls). This rhythm should be taken into consideration in the physiology of work and also for therapeutic purposes.

The presentation of the daily periodic course of latent vitality in the organism in relation to local time (fig. 1), shows the change from a sympathicotonic and a parasympathicotonic reaction situation. The maximum latent vitality is present at the change from the sympathicotonic to the parasympathicotonic phase and conversely. The change from the sympathicotonic to the parasympathicotonic reaction situation normally takes place between 9 a.m. and 10 a.m. and again between 8 p.m. and 9 p.m.; from the parasympathicotonic to the sympathicotonic reaction situation at about 3 a.m. and 3 p.m.

The extent of the sympathetic reaction can be increased by the intake of food or dextrose, physical movement, light-stimuli, the effects of stress, O_2-inhalation and by physical and other non-specific and specific measures, so that the change to the parasympathetic reaction phase then takes place at a higher level, in comparison with the spontaneuous course, and intensifying the metabolic changes.

The antithesis of sympathicotonia and parasympathicotonia has been described by *Hoff* [2] in a diagram of vegetative regulations with two

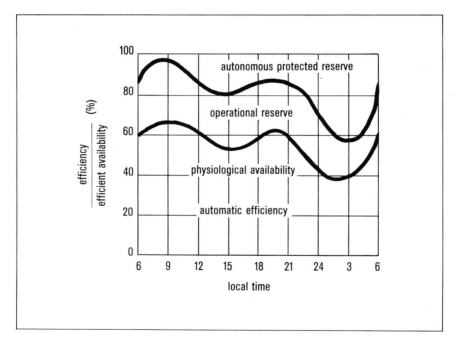

Fig. 1. Circadian rhythm: daily periodic course of the latent vitality (according to [15]).

phases of the total vegetative change-over (fig. 2). This affects mineral substances, acid-base balance, chemical, physicochemical and metabolic processes, and the endocrine system.

A non-specific stimulus enhances the course of reaction of the biological system in the same direction and possibly prolongs it in time if applied at the beginning of a reaction phase. If the stimulus is applied at the end of

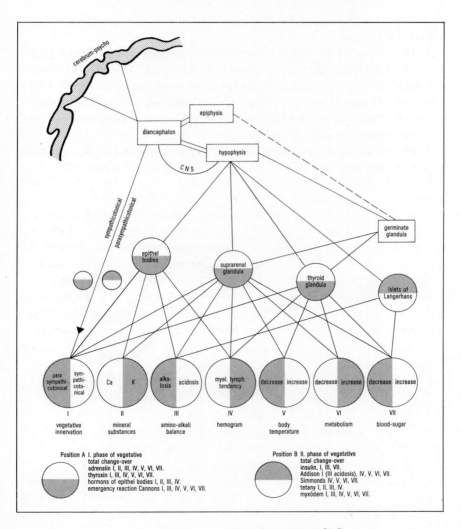

Fig. 2. Diagram of the vegetative regulations (according to [21]).

the sympathicotonic period, the parasympathicotonic phase is enhanced and prolonged. Contrary influences of the reaction course in the direction of the stimulus applied are likewise possible, e.g, changes in a parasympathicotonic reaction situation in the case of treatment by sympathicomimetics, toxins or stress stimuli.

Dependence of the Reaction on Stimuli of the Circadian Phase

The term "stress" coined by *Selye* [10] applies to a general reaction pattern in animals and man as a response to increased strain. Stresses can be, for example, of a physical (light, cold, heat, noise) or chemical (contaminants, drugs) nature, or non-specific effects as a result of medications, or of a mental nature (isolation, examinations, family burdens). The initial alarm phase in the form of a parasympathicotonic shock-reaction leads to the adaptation or counter-shock-phase with an increase in the tone of the sympathetic nervous system and hyperactivity of the adrenals, both of the medulla and of the cortex, plus a shrinking of the thymus and lymph nodes.

A certain degree of stress is not dangerous and is an essential part of life (eu-stress). However, severe stress over a long period (dys-stress) can cause damage to health in a variety of ways. Stomach ulcers, high blood-pressure and cardiac infarctions frequently result. Through the training of repeated subliminal stimuli, adaptation of the organism is achieved, so that even fairly severe stimuli do not have a pathogenic effect (eu-stress). The adaptability varies from one individual to another [12] and is subject to circadian rhythms. The effect of stimuli is more marked when occuring with a sympathicotonic reaction situation than with a parasympathicotonic situation. With his work on the total vegetative change-over, *Hoff* was one of the pioneers of biological cybernetics via feedback mechanisms [11, 16]. Regulation for the maintenance of homeostasis is a basic principle of the vital processes. However, the various feedback mechanisms are cross-linked and coupled with each other, in some cases in a multiple manner, and are thus reciprocally dependent so that changes in one parameter are transmitted to others, as is evident from the description of *Hoff*'s system of mechanisms (fig. 2). The control magnitude usually has the opposite sign to the deviation in the functional respect, i.e., when a parameter declines, this triggers an increase in it and vice-versa (negative feedback) (fig. 3).

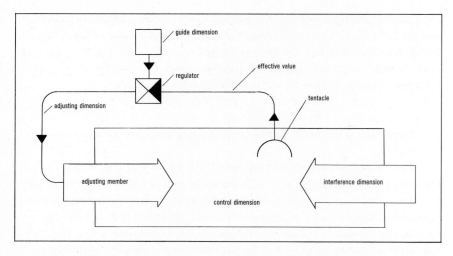

Fig. 3. Feedback mechanism in a representation appropriate for biological cybernetics. The technical control expressions are characterized by capital letters (according to [9]).

Vegetative Alteration by Non-Specific Stimuli

The ability to respond to stimuli is a basic characteristic of living systems. An alteration in the reaction situation of an organism can result from *any change* outside (external stimuli) or within (organ stimulus) causing an excitation or a reaction. Accordingly, for a cell, any influence is a stimulus which changes its *metabolic situation* and triggers counter-reactions. Either a metabolic activation or restitutio ad integrum then occurs or, if there is no adaptation, pathological reactions result. This clarifies initial deteriorations and late reactions in physical therapy and in homeopathy. Even placebo effects are subject to these natural laws.

Large doses of drugs appear to paralyze the reactivity of the cells. *Abrupt changes* in the vegetative reaction situation necessitate an adaptation. This has also been designated as a "massive-dose system treatment". How the cells and tissues in question react, depends, therefore, on the nature, intensity and duration of the changes in milieu, i.e., environment. Obviously, a role is also played by the genetic predisposition and topical disposition. With medication an almost 100% difference in effect was found when they were administered during contrary courses of reaction [3, 4].

The reactivity of the organism is also dependent on the *relative bodyweight*. A slow weight-reduction, in the case of adiposity, enhances any therapeutic measure at all and can even make this measure unnecessary.

Localized foci of inflammation are a persistent stimulus for the organism. In addition to exercising an effect on the immunological system and affecting the regulations via the active mesenchyma, they irritate the complex mechanisms of the total vegetative alteration and of the circadian rhythm. This is why the cleaning-up or inactivation of the foci is often a precondition for a cure.

Psychosomatic illnesses with organ manifestations often result through irritation of the vegetative regulations and of the circadian rhythm. This also applies to psychoses and addictions. Stimuli which promote and support the normal circadian rhythm, can be of as great a therapeutic value as systematic relaxation exercises through autogenous training and a habituation to mental stresses [3, 6, 8]. With an overstrained sympathicomimetic reaction situation, calming measures have a favorable effect and in the parasympathicomimetic phase, activating measures are appropriate.

Non-Specific Effect Even with Medicaments, Having a Specific Action

Non-specific measures and placebos have an effect on mental and physical mechanisms in a relatively large number of cases. It is, therefore, self-evident that therapeutic measures having a specific action, such as medical applications of a chemotherapeutic, vegetable, and organotropic nature with an organ-specific effect, especially biological mediators and carrier substances and immune factors and hormones, have an additional non-specific active component. This can play a major part in determining the biological or therapeutic effect by contributing to the recovery of the vegetative regulation capacity and to the synchronization of natural rhythms.

Summary

The daily phases of reaction situations, metabolism, and hormone regulation are presented in view of the total vegetative change-over with predominant sympathicotonus and, on the other hand, in view of the parasympathicotonus and of the many integration of vegetative regulations. Resulting conclusions give optimal timely therapy application in various illnesses.

Zusammenfassung

Die Tagesperiodik von Reaktionslage, Stoffwechsel und hormonaler Regulation werden unter dem Gesichtspunkt der vegetativen Gesamtumschaltung mit überwiegendem Sympathikotonus und andererseits des Parasympathikotonus unter der vielfältigen Verflechtung vegetativer Regulationen dargestellt. Daraus resultieren Konsequenzen für die optimale zeitliche Anwendung einer Therapie bei verschiedenen Erkrankungen.

References

1. Busse, E.: Akupunkturfibel (Pflaume, München 1975).
2. Hoff, F.: Lehrbuch der speziellen pathologischen Physiologie (Fischer, Jena 1945).
3. Kunkel, H.: Einflüsse der Biorhythmik – Stress und Nervensystem. Ärztl. Prax., Wien 28: 1803 (1975).
4. Lemmer, B.: Chronopharmakologie (Wissenschaftl. Verlagsges., Stuttgart 1983).
5. v. Mayersbach, H.: Zirkadiane Biorhythmik. Tagungsbericht XXIII. Jahrestagung über die Zytoplasmatische Therapie und die Methoden der Serum-Desensibilisierung, Stuttgart 1977.
6. Payk, T. R.: Neurol. Psychiat. 7: 363 (1980).
7. Pischinger, A.: Krebsarzt 21: 297 (1968).
8. Pohl, H.: Fortschr. Med. 85: 7 (1967).
9. Psychyrembel, W.: Klinisches Wörterbuch, p. 1031 (De Gruyter, Berlin-New York 1972).
10. Selye, H.: Streß beherrscht unser Leben (Econ, Düsseldorf 1957).
11. Shannon: The mathematical theory of communication (Urbana, 1959).
12. Theurer, K.: Medsche Klin. 60: 1909 (1965).
13. Theurer, K.: Ärztl. Prax., Wien 33: 1709 (1981).
14. Theurer, K.: Therapiewoche 33: 17 (1983).
15. Thiele, G.: Handlexikon der Medizin, p. 2734 (Urban & Schwarzenberg, München-Wien-Baltimore 1980).
16. Wiener, N.: Cybernetics (New York, 1961).

Prof. Dr. Karl E. Theurer, Forschungslaboratorien für Organo- und Immunotherapie, Brunnwiesenstraße 21, D-7302 Ostfildern-Ruit (FRG)

Therapeutic Integration of Xenogenic Proteins and Peptides in Modern Oncotherapy

H. Porcher

Stuttgart, FRG

The Possibilities and Limitations of Surgery, Radiotherapy, and Chemotherapy

The great majority of tumor cures are achieved by operative medicine. The success rate of a localized intervention, good operability being assumed, is between 40 and 70%. Accordingly, the aim of an interdisciplinary concept must be to totally eliminate the tumor cells still remaining in the body after an apparently radical operation which, time and again, can be the source of metastases and recurrences. In the final analysis, this can only be achieved by the activation of the body's own powers of resistance.

With the aid of regional radiotherapy, a cancer cure-rate of about 10% can be achieved. It is applied as a sole form of therapy or in combination with other forms. When used correctly, a therapy which is comparatively free from side-effects can certainly be assumed. However, effects on the hematogenic system are unavoidable, this being recognizable, for example, from an immunodepressive effect on T-lymphocytes, monocytes and O-cells. These effects can occur immediately after radiation and persist for some months or even years in some cases. Consequently, the purpose of a supportive measure to eliminate the bone-marrow suppression caused by radiotherapy must be the restoration of the immunological system.

Systemic chemotherapy covers the entire body and alone is capable of reaching even scattered micrometastases. About 10% of the forms of cancer can be healed by chemotherapy, especially acute lymphatic leukemia in children, Hodgkin's disease, chorion carcinoma in women, testicular cancer and so on (s. table I). The side-effects and the not inconsiderable proportion of chemotherapy-resistant tumors restrict the further range of application or necessitate, for the future, a therapy which makes chemotherapeutic agents more selective or their side-effects more tolerable.

The real therapeutic problem, however, is the weakening of the body's own powers of resistance. This weakening can initially be due to the tumor, but can be further accentuated by sequelae and by drastic

Table I. Classification of malignant tumors and system diseases by the success of chemotherapy [*DeVita*, 1975]

Group I
Tumor conditions potentially capable of responding to chemotherapy:

Acute infant lymphoblastic leukemia	Ewing sarcoma
Lymphogranulomatosis	Wilms tumor
Histiocyte lymphoma	Neuroblastoma
Testicular carcinoma	Burkitt lymphoma
Embryonal rhabdomyosarcoma	Retinoblastoma
	Chorion carcinoma in women

Group II
Tumor conditions with high remission rates under chemotherapy and prolongation of the survival time of patients:

Ovarian carcinoma	Lymphocytic lymphoma
Mammary carcinoma	Neuroblastoma
Acute leukemia in adults	Adrenal cortex carcinoma
Plasmacytoma	Malignant insulinoma
Endometrium carcinoma	Small-cell bronchial carcinoma
Prostatic carcinoma	Osteosarcoma

Group III
Tumors sensitive to chemotherapy, but without an appreciable prolongation of the survival time of patients:

Carcinomas of the head and neck region	Malignant melanoma
Gastrointestinal carcinomas	Malignant carcinoid
CNS tumors	Soft-tissue sarcoma
Endocrine gland tumor	Bladder carcinoma

Group IV
Largely resistant tumors:

Hypernephroma	Pancreas carcinoma
Oesophagus tumor	Liver carcinoma
Epithelioma of the bronchus	Thyroid carcinoma

therapeutic measures. Only when this is realized, is it also understandable why radical and super-radical operative techniques have not resulted in the perfect tumor cure anticipated. The precondition for better control of cancer is, therefore, a fundamental change in the structure of the methods used for combatting cancer up till now. Interdisciplinary collaboration and new cancer strategies are needed.

Biological Response Modifier: Physiological Approaches in Oncotherapy

The medicamentous stimulation of the body's own defences is becoming increasingly important in tumor therapy. In English-speaking countries, substances which are capable of stimulating the body's own powers of resistance are described as "biological response modifiers". These are defined as "substances or measures which modify the relation between tumors and host by modulation of the biological reaction of the host to its tumor with a resultant therapeutic effect" (from the news-sheet of the National Cancer Institute, Bethesda, Md., USA, June 1981).

Numerous mediators are concerned in the biological reaction of the host to its tumor. These include, for example, lymphokines, interferons, prostaglandins, coagulation factors, etc. These are therefore not exogenous chemicals, cytostatic drugs or therapeutic agents in the conventional sense, but endogenous biological mediators which are involved in the biological reaction anyway. Immunomodulatory mechanisms and the direct regulation of tumor-cell metabolism within the complex carcinoma-host relation will be at the center of research in new approaches to cancer therapy.

Immunomodulatory and Antitumoral Action of Xenogenic Peptides

A whole series of peptides and proteins from xenogenic tissues (liver, spleen, decidua, thymus) are capable of inhibiting human tumor cells in cultures in their synthesis activity, while normal cells display a tendency to cell stimulation. These factors thus inhibit the DNA synthesis-rate of various tumor-cell lines (Wish, HEP II, melanoma) in a dose-related manner (fig. 1). In contrast to this, normal cells remain uninfluenced in their growth behavior or are even stimulated. This marked effect

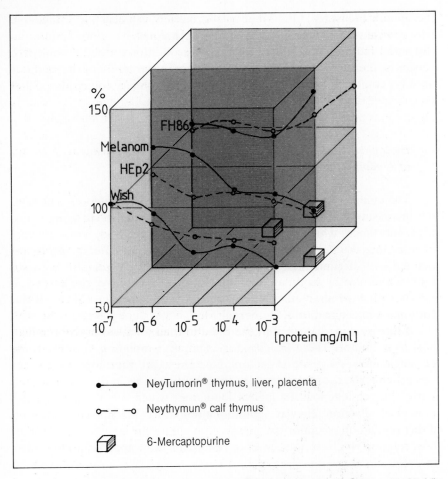

Fig. 1. Dose-effect relation of NeyTumorin®-Sol and Neythymun® (after *T. Stiefel*) with various malignant (Wish, HEP 2, melanoma) and benign (FH 86) human cells. The incorporation rate of tritium-labeled thymidine in the DNA was measured. The 100% value corresponds to the incorporation rate of control cultures which were treated only with the solvent (physiological saline solution). The inhibition values of the purine antagonist 6-mercaptopurine are entered for comparison.

on tumor cells has been demonstrated in numerous tumor- and normal-cell cultures [1–4, 8]. Interestingly enough, tumor-cell membranes dispose of a higher receptor density than normal cells, which may explain the higher affinity of these factors for tumor cells [5].

It has been demonstrated in animal experimentation, that not only the growth of syngenic tumors, but also already established tumors, can be regressed in 50–80% of cases [6]. These effects can likewise be explained by an activation of the immune resistance, since they do not occur with animal variants not having a thymus, i.e., also by a direct effect in the nature of an inhibition of the DNA synthesis [7]. It is assumed that these active agents are low- to medium-molecular fractions of proteins or polypeptides. Spleen cell cultures of mice stimulated with xenogenic peptides led to a mobilization of cytotoxic T-lymphocytes against syngenic tumors, almost all the tumor cells in the culture being destroyed within 1–2 days [9].

Radiation-Protective Action of Xenogenic Peptides at LD 50

In addition to these direct antitumoral effects and indirect effects demonstrated by the immunological system, peptide mixtures (NeyTumorin®*) have been shown to provide a marked radiation protection. If experimental animals are exposed to a dose of radiation, which is lethal for 50% of the group of animals (LD_{50}), and the animals are treated with these factors after exposure to the radiation, all animals survive the high radiation exposure (fig. 2). In connection with radiation therapy, these substances should be of clinical significance with respect, among other things, to a reduction of possible side-effects.

Significance of Xenogenic Peptides in Oncology

Xenogenic proteins and peptides will not be able to supersede operations, radiation treatment or chemotherapy, but they are a necessary supplement, especially with respect to the reduction of therapy-related side-effects and to increasing the body's own powers of resistance. As in the past, a clinically manifest tumor must be reduced to the greatest possible extent by surgical, radiation or chemocytostatic measures. Xenogenic proteins and peptides will, therefore, in future assume the position they merit in oncology in the reduction of the side-effects of chemotherapy (table II) and radiation treatment, in the treatment of inoperable tumors and in improving the quality of life.

* vitOrgan Arzneimittel GmbH, D-7302 Ostfildern 1, FRG

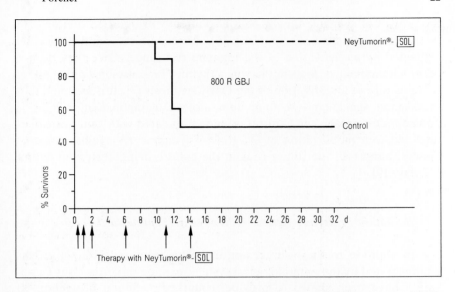

Fig. 2. Cytoprotective effect of NeyTumorin®-Sol in mice exposed to the LD_{50} (800 R whole-body irradiation). The treatment with NeyTumorin®-Sol enables 100% of the group treated to survive the radiation damage.

Table II. Indications of xenogenic proteins and peptides in combination with chemotherapy

(1) Malignant growths of the DeVita group II to IV.
(2) Malignant growths of the DeVita group I and II when the patients have been treated with all the conventional therapeutic methods.
(3) Malignant growths of the DeVita group I to IV when a patient refuses one or all of the therapeutic methods proposed.

Through the combination, the tolerance of chemotherapy is substantially improved and the side-effect rate – especially in respect to acute side-effects – is significantly lowered. A simultaneous application is recommended: bypass for infusion of cytostatic agents.

Summary

The present-day therapeutic concept for tumors essentially comprises surgery, radiotherapy, and chemotherapy. Supportive and symptomatic measures include infection prophylaxis for the avoidance of complications of chemotherapy, intensive measures for controlling the toxic side-effects of chemotherapy, nutritional control, alleviation of pain and medical care for the emotional problems of cancer patients and rehabilitation. Efforts have recently been made to also include immunotherapy, at least as a supportive measure, in the tumor therapy concept.

Zusammenfassung

Das heutige Konzept für die Therapie von Geschwülsten besteht im wesentlichen aus Chirurgie, Radiotherapie und Chemotherapie. Zu den unterstützenden und symptomatischen Maßnahmen gehören Infektionsprophylaxe zur Verhütung von Komplikationen der Chemotherapie, Intensivbekämpfung der toxischen Nebenwirkungen der Chemotherapie, Sorge für Ernährung, Schmerzlinderung sowie psychische Betreuung der Krebskranken und Rehabilitation. Neuerdings sind Bestrebungen im Gange, die Immunotherapie, zumindest als supportive Maßnahme, in das tumortherapeutische Konzept mit einzubeziehen.

References

1. Ketelsen, U.-P.: Pilotstudie zum Einfluß eines biologischen "Response Modifiers" (NeyTumorin®) auf die Plasmamembran menschlicher Tumorzellen (Wish) in vitro im Vergleich zu einem Chemozytostatikum (6-Mercaptopurin). Therapiewoche 33: 62–70 (1983).
2. Letnansky, K.: Stoffwechselregulatoren der Plazenta und ihre Wirkung in Normal- und Tumorzellen. Exp. Pathol. 8: 205–212 (1973).
3. Letnansky, K.: Tumorspezifische Faktoren der Plazenta und Zellproliferation. Exp. Pathol. 9: 354–360 (1974).
4. Letnansky, K.: Inhibition of thymidine incorporation into the DNA of normal and neoplastic cells by a factor from bovine maternal placenta: interaction of the inhibitor with cell membranes. Biosci. Rep. 2: 39–45 (1982).
5. Letnansky, K.: Entdeckung zellulärer Rezeptoren für antitumorale plazentare Faktoren in NeyTumorin. Therapiewoche 33: 59–61 (1983).
6. Munder, P. G.; Stiefel, T.; Widmann, K. H.; Theurer, K.: Antitumorale Wirkung xenogener Substanzen in vivo und in vitro. Onkologie 5: 2–7 (1982).
7. Munder, P. G.: Experimentelle Untersuchungen über den antitumoralen Wirkungsmechanismus von NeyTumorin. Therapiewoche 33: 71–73 (1983).
8. Theurer, K.; Paffenholz, V.: Einfluß von makromolekularen Organsubstanzen auf menschliche Zellen in vitro. I. Diploide Kulturen. Kassenarzt 27: 5218–5226 (1978). II. Tumorzellkulturen. Kassenarzt 19: 1876–1887 (1979).
9. Report on the XIIIth International Cancer Congress in Seattle, Washington (USA), Sept. 1982. Heilkunst 4 (1983).

Dr. H. Porcher, Kernerstr. 26, D-7000 Stuttgart 1 (FRG)

Experimental Studies

Study of the Efficacy of Organic Extracts on a Cell-Free System of Hela Cells*

D. Jachertz[1], B. Jachertz[1], G. Mai[2]

[1] Institute for Therapeutic Biochemistry of the Univesity of Frankfurt/Main, FRG
[2] Hygiene-Institute of the City and University of Frankfurt/Main, FRG

Introduction**

The efficacy of antibiotics, pharmaceutical preparations and also therapeutically applied organic extracts, can be studied not only in experimental animals and cell cultures, but also in cell-free biosynthetic systems [14]. The investigation of such substances in cell-free systems has the advantage that on one hand, it is possible to accurately measure the biological activity and, on the other hand, that the mechanism of action of the drugs studied can be clarified in favorable cases.

Before the results of the study of organic extracts in our cell-free system are described, the theoretical and experimental fundamentals of biosynthetic systems should be set out.

Material and Methods

The cell-free biosynthetic system was prepared from Hela cells as described by *Nirenberg et al.* [12]. It contained the substances listed in table I, with the exception of the messenger-RNA.

The Hela cells were bred in continuous culture by *May*'s method [13].

---

* Medsche Klin. *18:* 752–754 (1963).

** The essential results of this paper have obliged us to republish this manuscript in the present issue. The guiding impulses of this publication stimulated essentially the development of the cytoplasmatic therapy.

For the synthesis of protein, the following components are needed in cells of macro- and microorganisms:

(1) DNA as information constituting the blue-print of the protein to be synthesized;

(2) a synthesizing apparatus for the conversion of the information;

(3) elements which as energy-rich compounds permit the synthesis to take place in an energetically exothermic manner.

The *information* is deposited in all cells in the form of desoxyribonucleic acid (DNA) [5]. It can be compared with a strip of paper on which Morse signals are written. The carrier (corresponding to the strip of paper) is a phosphoric acid-chain incorporating esterlike branches of desoxyribose. These carry four different nucleic bases (comparable to the Morse signals) which, probably in triplet form [7], each represent a letter. For biological, chemical and physical reasons, which will not be discussed in more detail here, the DNA is usually present as a double stranded molecule, corresponding to two strips of paper, wrapped around each other, with Morse signals. In this, an adenine in one strand always has as its counterpart a thymine in another; a cytosine in one strand always has a guanine as its opposite number in another [6]. If a base [1] is taken from the natural sequence of the nucleic bases or if one base is replaced by another base [21] or by a base-analogue compound [10], a mutation or the loss of vitality is the consequence, since the information at this particular point has lost its biological sense.

The *synthesizing apparatus* is of a more complicated structure. It has been known for some time that the synthesis of proteins is usually structure-linked [9]. These synthesizing structures in cells of macroorganisms are the microsomes and in microorganisms the ribosomes. They are small particles consisting of 63% of ribonucleic acid (RNA) and up to 37% of protein [4]. This structure must now be supplied; on the one hand, with information from the DNA and, on the other hand, with structural material from the cytoplasm [18]. In addition, one or more enzymes must act on the structure supplied with information and elements to join up the elements arranged in accordance with the information, as a protein by peptide-like links [14].

The *information* in the cell can only be brought from the DNA to the synthesizing structure in a material form. The substance transmitting the information is a RNA, which is thus designated as a messenger-RNA (m-RNA) [2]. It is built as a complementary element to the sequence of bases in the DNA [8], just as one DNA-strand in the double strand is complementary to its partner. During the synthesis of the protein, the information in the form of the messenger-RNA is translated to the structure to be synthesized in a manner which is still unknown [2]. It has only a very short life [2].

The *structural material* consists of activated amino acids [9]. They are transported by RNA molecules which, in accordance with this function, are known as transfer-RNA [17]. This transfer-RNA transports the amino acids to the synthesizing structures and, after the synthesis has been completed, leaves them in a "discharged" state as it were – without amino acid – to bring forth new amino acid molecules again [2, 14].

The interplay of all these components is shown in the diagram (fig. 1).

The macromolecules participating in the protein synthesis are present in the cell-free biosynthetic system in the same way as they are in the cell, but they are subject to experimental changes or are omitted from the system to allow their natural function to be identified. Cell-free biosynthetic systems are thus put together by the investigator from individual parts obtained from appropriate cells. The DNA as the carrier of the information can be obtained from cells by various preparative techniques [11]. It can be removed from extracts by digestion with desoxyribonuclease [12].

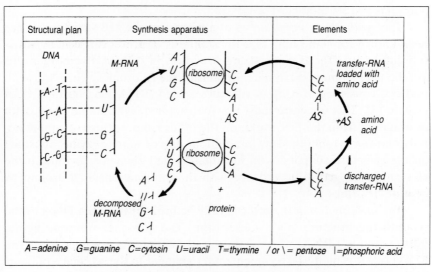

Fig. 1. Components for the synthesis of proteins.

As a rule, the ribosomes or microsomes are obtained from cell extracts by centrifuging and can be sedimented on high-speed centrifuges (100 000 g) in different gradients [14]. They are available with greater purity. The transfer-RNA for use in cell-free biosynthetic systems is prepared by sedimentalian procedures [14]. Together with proteins and low-molecular substances, it is contained in the supernatant after centrifuging at 100 000 g. Dialysis of the 100 000 supernatant removes the low-molecular components and leaves behind 20 different transfer-RNA molecules, 20 different enzymes activating the amino acids and positioned attaching them at the transfer-RNA (one for each amino acid) and a still unknown number of peptide-binding enzymes [14].

The messenger-RNA decays and, therefore, does not appear in the 100 000 g supernatant. It is usually artificially added after meticulous preparation or as biologically senseless information in the form of polymerisates of nucleoside diphosphate prepared in vitro [15].

The amino acids are not added to the cell-free systems in the activated form. They are activated by the corresponding enzymes contained in the dialyzed 100 000 g supernatant, assuming that an ATP-generating system censures the necessary supply of energy-rich bonds.

A biosynthetic system is composed by components as shown in table I.

Material and Method – Our Own Studies

The cell-free biosynthetic system was prepared from Hela cells as described by *Nirenberg et al.* [12]. It contained the substances listed in table I with the exeption of the messenger-RNA. The Hela cells were bred in continuous culture by *May's* method [13].

Table I. Composition of a biosynthetic system

Washed ribosomes 100 000 g supernatant, dialyzed messenger-RNA (possibly polymerisate) tris-buffer, pH 7.8	0.1 m
Magnesium acetate	0.01 m
Potassium chloride	0.005 m
Mercapto ethanol	0.006 m
ATP	0.001 m
Phosphorenol pyruvate	0.005 m
Pyruvate kinase	20.0 y/ml
GTP	0.03 m
Amino acids, one of which is radioactive	0.000 005 m each

Organic extracts were obtained according to *Theurer*'s method [19], from the organs of healthy animals* and subjected to vacuum hydrolysis [20].

The protein synthesis was observed by the incorporation of radioactive phenylalanine (3-^{14}C), specific activity 1 mc/mM, in the protein precipitable in 5% TCA [16].

The measurement of the radioactivity was carried out in a Tricarb scintillalian counter.

The time of incubation for synthesis is shown in the tables.

Results

First of all, we checked whether organic extracts were capable of starting the protein synthesis in the cell-free biosynthetic system of Hela cells. This system lacked the messenger-RNA and the DNA so that by itself it was not able to synthesize any quantities of protein during the incubation period. As shown in table II, it is possible with some organic extracts to initiate protein synthesis in the cell-free system in clearly detectable amounts. The differences in the protein synthesis achieved with the individual extracts are discussed later.

After other studies with a cell-free biosynthetic system of Escherichia coli, it became known that the information for the phenylalanine incorporation i.e., the corresponding messenger-RNA, was sensitive to UV light, but that the corresponding transfer-RNA was UV-resistant and finally, that the corresponding places at the ribosomes were again sensitive to UV

* The extracts were obtained from the organs of the following healthy animals:

Organ	Animal
Foetal placenta	Cattle
Maternal placenta	Cattle
Mucosa	Pigs
Kidney	Cattle (calf)
Lung	Cattle, foetal
Bone-marrow	Cattle
Liver	Pigs

Table II. Phenylalanine (3-^{14}C)-incorporation after completion of biosynthetic system by addition of extracts of different organs (Ipm = impulses/min)

Organic extract	Incorporated phenylalanine (Ipm)
System without extract	0
with extract of foetal placenta	2356
with extract of maternal placenta	0
with extract of mucosae	789
with extract of kidney	1044
with extract of lung	1643
with extract of bone-marrow	0
with extract of liver	0
Duration of test 30 min	

light [23]. We attempted to halt the protein synthesis in our cell-free system, too, by irradiation of the microsomes (which correspond to the ribosomes) with UV light. As shown in table III, this damage can be repaired by extract of foetal placenta.

On the assumption that the information of the extract used for the protein synthesis is determined by its content of biologically active DNA, we digested the extract with DNase. We then expected a decrease in protein synthesis. As the result in table IV shows, the opposite is the case.

Instead of a digestion with RNase which would have destroyed the cell-free system, we endeavored to demonstrate the biological and intact function of the transfer-RNA in the organic extract used in an indirect

Table III. Phenylalanine (3-^{14}C)-incorporation after completion of biosynthetic system damaged by UV radiation by addition of extract of foetal placenta

Test-conditions	Incorporated phenylalanine (Ipm)
System without extract (control)	401
System, microsomes irradiated with 3.5 10^4 erg/mm^2 UV	12
System with microsomes as described above after addition of extract of foetal placenta	840
System 100 000 g supernatant irradiated with 3.5 10^4 erg/mm^2 UV	930
System with microsomes irradiated as above and irradiated 100 000 g supernatant after addition of extract of foetal placenta	960
Duration of test 30 min	

manner by destroying this function with UV light. As is apparent from table IV, the effect of the UV exposure of the extract is moderate.

Finally, we digested the extract with trypsin, which was inactivated after the end of the digestion by briefly heating to 100° C. After the tryptic digestion, the extract had practically lost its biological function, as is apparent from the result shown in table IV.

Surprising was the finding that extract of foetal placenta and also of other organs still displayed a good efficacy in our biosynthetic system, even in very high dilutions (table IV).

Discussion

The results demonstrate that a surprisingly good biological activity of organic extracts tested can be shown. This initially appears apparent in a certain organ specificity, even though it cannot be detected in which macromolecular instance the specificity is determined (see table II).

The DNA contained in the extracts can very well still be biologically active. In our study, it could have a regulatory function and maintain the protein synthesis at a certain level. If this regulatory function is eliminated, protein is synthesized in an irregular and excessive manner.

The relative insensitivity of the RNA in the extracts to the UV irradiation cannot be clearly explained. Either they are present in excess quantity in the extracts, or they play no role in our system for protein

Table IV. Phenylalanine (3-^{14}C)-incorporation after completion of biosynthetic system by various pretreated extracts of foetal placenta

Preliminary treatment	Incorporated phenylalanine (Ipm)
System without extract (control)	0
+ polyuridyl acid as messenger-RNA	3657
+ extract of foetal placenta	1079
+ extract of foetal placenta diluted to 0.1 y/ml	627
+ extract of foetal placenta after DNase digestion	2441
+ extract of foetal placenta after trypsin digestion	255
+ extract of foetal placenta after UV irradiation	935

synthesis. This can be assumed for the messenger-RNA but is uncertain for the transfer- and microsomale-RNA. We consider it likely that the extract's DNA in our system is an information-carrier. The formation of a corresponding messenger-RNA is initiated either by this substance or by the system.

The effect of the extract's protein components is particularly interesting. This is, above all, to be sought in the most diverse synthesizing enzyme activities, when utilizing what is already known as a guide. It can, however, be located in mechanisms which are still totally unknown.

Although the extract's mechanism of action studied could not be attributed to one or other known mechanisms involved in protein synthesis, it is, nevertheless, possible to convincingly demonstrate the actual efficacy of such extracts. From the present results, it is most likely that the efficacy of the extracts is based on several factors.

Finally, if the entire content of the cells is passed by the extracts to the biosynthetic system from the wide range available, most use is made for synthesis purposes of that substance, which the system is lacking most.

Summary

The biological activity of organic extracts was studied in a cell-free biosynthetic system. It was shown that for a specific cell-free system only distinct extracts can develop an efficacy. The efficacy of these extracts is based on diverse mechanisms which are all supplied to the system in a biologically active form.

Zusammenfassung

In einem zellfreien biosynthetischen System wurde die biologische Aktivität von Organextrakten geprüft. Dabei zeigte sich, daß für ein bestimmtes zellfreies System nur ganz bestimmte Extrakte eine Wirksamkeit entfalten können. Die Wirksamkeit dieser Extrakte ist auf vielseitige Mechanismen gegründet, die alle in biologisch aktiver Form dem System zugeführt werden.

References

1 Bautz, E.; Freese, E.: Proc. natn. Acad. Sci. USA *46:* 1585 (1960).
  Crick, F. H. C.; Barnett, L.; Brenner, S.; Watts-Tobin, R. J.: Nature *192:* 1227 (1961).
2 Brenner, S.; Jacob, F.; Meselson, M.: Nature *190:* 576 (1961).
3 Berg, P.: Ann. Rev. Biochem. *30:* 293 (1961).
4 Brown, G. L.: Br. med. Bull. *18:* 10 (1962).

5   Chemie der Genetik, 9. Colloquium der Ges. für Physiol. Chemie, 17.–19. 4. 1958, Mosbach/Baden.
6   Crick, F. H. C.; Barnett, L.; Brenner, S.; Watts-Tobin, R. J.: Nature *192:* 1227 (1961).
7   Gamov, G.: Biol. Medd. Dan. Vid. Selsk *22:* 2 (1954).
8   Haiashi, M.; Spiegelmann, S.: Proc. natn. Acad. Sci. USA *47:* 1564 (1961).
9   Hoagland, M. B.: In Chargaff, Davidson (eds.), Nucleic Acids III, p. 349; Gros, F., p. 409 (Academic Press, New York 1960).
10 Litman, R. M.; Pardee, A. B.: Nature *178:* 529 (1956).
    Weygand, F.; Wacker, A.; Dellweg, H. W.: Z. Naturf. *7 b:* 19 (1952).
    Freese, E.: J. molec. Biol. *1:* 87 (1959).
11 Marmur, J.: J. molec. Biol. *3:* 208 (1961).
12 Matthaei, J. H.; Nirenberg, M. W.: Proc. natn. Acad. Sci. USA *47:* 1580 (1961).
    Nathans, D.; Lipmann, F.: Proc. natn. Acad. Sci. USA *47:* 497 (1961).
13 May, G. (in press).
14 Nathans, D.; Lipmann, F.: Proc. natn. Acad. Sci. USA *47:* 497 (1961).
15 Nirenberg, M. W.; Mathaei, J. H.: Proc. natn. Acad. Sci. USA *47:* 1588 (1961).
16 Novelli, D.; Mans, R. J.: Arch. Biochem. Biophys. (in press).
17 Schweet, R.; Bovard, F. C.; Allen, E.; Glassmann, E.: Proc. natn. Acad. Sci. USA *44:* 173 (1958).
18 Symposium on Amino Acid Activation. Proc. natn. Acad. Sci. USA *44:* 67 (1958).
19 Theurer, K.: Ärztl. Praxis *11:* 1120, 1167 (1959).
20 Theurer, K.: Verfahren zum gesteuerten chemischen Aufschluß von organischen Stoffen und biologischen Geweben für therapeutische Zwecke. DBP 1 090 821.
21 Vielmetter, W.; Schuster, H.: Z. Naturf. *15 b:* 298 (1960).
22 Watson, J. D.; Crick, F. H. C.: Nature *171:* 737 (1953).
    Cold Spring Harbor Symposium. Q. Biol. *18:* 123 (1953).
23 Wacker, A.; Jachertz, D.; Jachertz, B.: Angew. Chem. *74:* 653 (1962).

Prof. Dr. med. D. Jachertz, Direktor des Instituts für Hygiene der Mikrobiologie der Universität Bern, Friedbühlstr. 51, CH-3010 Bern (Schweiz)

Synthesis of Thymopoietin 32-36 (TP 5) and its Effect on the Growth of Fibrosarcoma*

W. Voelter[1], Ch. Friedl[1], H. Kalbacher[1], P. G. Munder[2], K. H. Widmann[2]

[1] Dept. of Physical Biochemistry, Physiological-Chemical Institute of Tübingen University, Tübingen, FRG
[2] Max Planck Institute for Immunobiology, Freiburg-Zähringen, FRG

The hormonal effects of thymus polypeptides on the immune-defence system have, for a considerable time, been intensively studied [1–5]. Thymopoietin II, a polypeptide consisting of 49 amino acids, was isolated for the first time by *Goldstein* [6] in May 1974 from calf thymus, the first proposed sequence of 1975 [7] being corrected by *Goldstein* in 1981 [8]. The search for smaller peptides with similar biological activity relating to the immune system as thymopoietin, led to the discovery of the pentapeptide thymopoietin 32-36 (TP 5). TP 5 was synthesized, both, by the classical method [9, 10], however, without any available details of the synthesis process, and also by the solid phase method [11]; see formula:

$$H_2N-CH-\overset{\overset{O}{\|}}{C}-NH-CH-\overset{\overset{O}{\|}}{C}-NH-CH-\overset{\overset{O}{\|}}{C}-NH-CH-\overset{\overset{O}{\|}}{C}-NH-CH-COOH$$

with side chains: $(CH_2)_3$–NH–C(=NH)–NH$_2$; $(CH_2)_4$–NH$_2$; CH_2–COOH; CH(CH$_3$)$_2$; CH$_2$–C$_6$H$_4$–OH

Amino Acid Sequence of TP 5 (Arg-Lys-Asp-Val-Tyr)

* The abbreviations for amino acids and protecting groups correspond to those of the IUPAC-IUE Nomenclature Commission: Eur. J. Biochem. 27: 201 (1972); 1 (1977).

To test the effect of TP 5 on fibrosarcoma induced in mice with methylcholanthrene, a classical synthesis method for the pentapeptide is described here.

The peptide is synthesized by successive linking of amino acids, and non-crystallizing intermediate compounds are crystallized as dicyclohexyl-ammonium(DCHA) salts. As shown in figure 1, all trifunctional amino acids are completely protected. Classical protective groups perform this function. The coupling methods used are the active ester method, and the dicyclohexylcarbodiimide (DCCI) method in an additive manner. The pentapeptide is finally released in two stages: all acid unstable protective groups are split off with trifluoroacetic acid (TFA), and then the benzyloxycarbonyl(Z) groups are removed by catalytic hydrogenation. The dipeptide Z-L-Asp(β-OBu^t)-L-Val-OH is produced without difficulty by means of active ester coupling. The preparation of pure H-Tyr(Bu^t)-OBu^t gives a 61%-yield, and coupling by means of DCCI gives readily crystallizing tripeptide Z-L-Asp(β-OBu^t)-L-Val-L-Tyr(Bu^t)-OBu^t. After release of the N-terminal amino groups by catalytic hydrogenation, extension to the tetrapeptide is performed by reaction of the tripeptide with Z-L-Lys(Boc)-OSu. Release of the terminal amino group provides the tetrapeptide H-L-Lys(Boc)-L-Asp(β-OBu^t)-L-Val-L-Tyr(Bu^t)-OBu^t which,

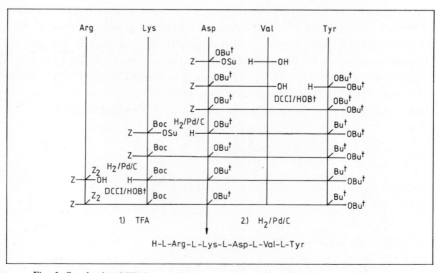

Fig. 1. Synthesis of TP 5.

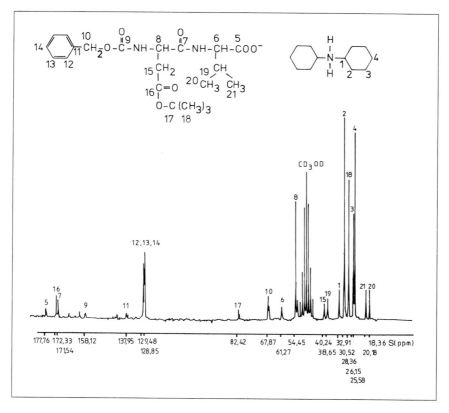

Fig. 2. ^{13}C-NMR spectrum of Z-L-Asp(β-OBu^ت)-L-Val-OH · DCHA proton broadband decoupled, 200 mg in 2 ml CD$_3$OD.

using DCCI/HOBt, is coupled with Z-L-Arg(Z$_2$)-OH to form Z-L-Arg(Z$_2$)-L-Lys(Boc)-L-Asp(β-OBu^ت)-L-Val-L-Tyr(Bu^ت)-OBu^t. The protecting groups are removed in the first stage within 90 min with anhydrous TFA, then the Z-groups are hydrogenated off by catalysis within 2 h. Subsequent gel filtration on Sephadex G 15 and lyophilization yields pure TP 5.

Purity control and characterization are performed with the aid of elementary analysis, amino acid analysis, racemate test, high-pressure liquid chromatography and ^{13}C-NMR spectroscopy. Figure 2 shows the ^{13}C-NMR spectrum of Z-L-Asp(β-OBu^t)-L-Val-OH · DCHA; figure 3, shows the HPLC chromatogram of released, purified TP 5.

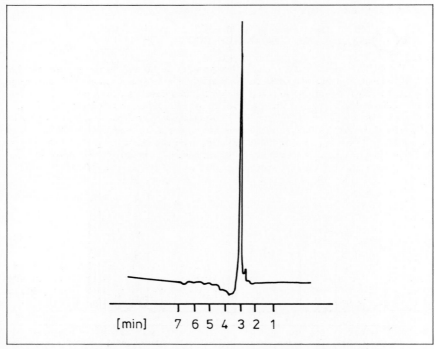

Fig. 3. HPLC chromatogram of thymopoietin (32-36). Column: 24 × 0.4 cm LiChrosorb RP-18. Eluent: $CH_3CN/0.01$ N NH_4Ac buffer, pH 4.0; 0.5 : 95.5. Detection at 278 nm.

The effect of TP 5 on the growth of a fibrosarcoma induced by methylcholanthrene (Meth.A.asc.) is tested in female CFB_1 (Balb/c ♀ C57 ♂) mice. Although the tumor growth is decisively reduced during the first 14 days, none of the animals survives longer than 28 days. Thymosin β9 (1–7) and the whole thymus extract show similar effects (cf. table I).

Table I. Influence of thymopoietin 32–36, thymosin β9(1–7), and whole thymus extract on the growth of the fibrosarcoma induced by methylcholanthrene

	Tumor volume in %				
	7	14	21	28	days after transplantation
Control	100	100	100	100	
TP 5	96	67	93	93	
Thymosin β9(1–7)	110	54	55	73	
Thymus extract	97	55	50	71	

Experimental Part

Synthesis

The melting points are determined in open capillaries on an apparatus according to Dr. Tottoli (Buechi, Switzerland), and are not corrected. The specific rotation values are measured with a Zeiss Old-5-Polarimeter. The amino acid analyses are performed on an automatic analyzer Liquimat III (Kontron, Eschingen, FRG). For this purpose, samples are hydrolyzed for 24 h with 6 N HCl at 120° C. The elementary analyses are carried out on an Analyzer 140 B (Perkin-Elmer). The percentage content of D-amino acids is determined by gas chromatography (apparatus Carlo Erba 3900) on the chiral phase of the N-(pentafluoropropionyl)amino acid propyl ester. All synthesized compounds are checked for uniformity by means of thin-layer chromatography, using TLC ready-made plates 60 F 254 (Merck, Darmstadt, FRG). The following elution systems are used:

A:	n-butanol/HAc/H_2O	3 : 1 : 1
D:	chloroform/MeOH/benzene/H_2O	8 : 8 : 8 : 1
E:	chloroform/MeOH	9 : 1
H:	chloroform/MeOH/HAc	90 : 8 : 2
J:	n-butanol/ethyl acetate/HAc/H_2O	1 : 1 : 1 : 1
K:	n-butanol/pyridine/HAcH_2O	5 : 5 : 1 : 4.

The substances are made visible by spraying with ninhydrin solution [12] and with chlorotolidine solution [13]. HPLC measurements are performed on the Jasco Twincle apparatus. The ^{13}C-NMR spectrum is recorded on a WP-80-NMR spectrometer (20.115 MHz, Bruker, Karlsruhe, FRG).

Z-L-Asp-OH

106.5 g (0.8 mol) L-aspartic acid are suspended in 400 ml H_2O/dioxane (3 : 1), then, 134.4 (1.6 mol) NaHCO$_3$ and 150 g (0.88 mol) chloroformic benzylester (Z-chloride) are added (dropwise) while stirring. During this procedure, the pH value is kept constant at 9–10 with 4 N NaOH, and the mixture is cooled to −5 to 0° C with ice/common salt mixture. When the reaction is completed, the mixture is filtered off from the undissolved NaHCO$_3$, and the filtrate is extracted with ether and cooled to 0° C; the aqueous phase is covered with a layer of ethyl acetate (EA) and carefully acidified with conc. HCl. The EA phase is separated and the aqueous phase extracted with EA, then the pooled EA phases are washed with 2 N citric acid and with saturated NaCl solution. After drying over Na$_2$SO$_4$, the phases are slightly constricted. To the solution light petroleum (PE) is added at low temperature, and Z-L-Asp-OH slowly crystallizes out.

Yield 160.2 g (75%); melting point 117° C; uniformly in the systems A, D and H; $[\alpha]_D^{20} = +9.3 \pm 0.5°$ C (c = 2, HAc).

$C_{12}H_{13}NO_6$ (267.24)
Calc.:	C 53.94	H 4.90	N 5.24
Found:	C 54.42	H 5.01	N 5.30

Z-L-$\overline{Asp\text{-}O}$

The anhydride is prepared from Z-L-Asp-OH according to ref. [14] with 76.6% yield; melting point 107° C; $[\alpha]_D^{20} = -36.2 \pm 0.5°$ (c = 2.25, HAc); uniformly in the systems A and D.

$C_{12}H_{11}NO_5$ (249.22)
Calc.:	C 57.93	H 4.45	N 5.62
Found:	C 56.15	H 4.75	N 5.18

Z-L-Asp-ONb

The ester is prepared according to ref. [15] with 83.5% yield; melting point 123–124° C; $[\alpha]_D^{22} = -17.1 \pm 0.5°$ (c = 1.95, HAc); uniformly in the systems A and D.
$C_{19}H_{18}N_2O_8$ (402.36)

Calc.:	C 56.71	H 4.51	N 6.96
Found:	C 56.12	H 4.64	N 6.53

Z-L-Asp(β-OBu^t)-ONb

The tert. butylation of Z-L-Asp-ONb is performed after suspension in CH_2Cl_2 with a large excess of isobutylene under sulphuric acid catalysis (2 ml H_2SO_4 in 600 ml CH_2Cl_2) giving 65% yield. Work-up is performed after a 5-day reaction time, according to [16]; melting point 94° C; $[\alpha]_D^{22} = -16.0 \pm 0.5°$; uniformly in the systems A and D.
$C_{23}H_{20}N_2O_8$ (458.47)

Calc.:	C 60.26	H 5.72	N 6.11
Found:	C 59.76	H 5.64	N 6.04

Z-L-Asp(β-OBu^t)-OH · DCHA

Z-L-Asp(β-OBu^t)-ONb is saponified to Z-L-Asp(β-OBu^t)-OH, according to [16], with 98% yield and then converted to the dicyclohexylammonium salt; melting point 124–125° C; $[\alpha]_D^{23} = +12.6 \pm 0.5°$; uniformly in D and H.
$C_{28}H_{44}N_2O_6$ (504.67)

Calc.:	C 66.64	H 8.79	N 5.55
Found:	C 66.78	H 8.76	N 5.51

Z-L-Asp(β-OBu^t)-OSu 1

12.8 g (39.7 mmol) Z-L-Asp(β-OBu^t)-ONb and 4.6 g (40 mmol) N-hydroxysuccinimide (HOSu) are dissolved in 40 ml dioxane and cooled to 0° C. At this temperature 8.95 g (40 mmol) DCCI, dissolved in 8 ml cold dioxane, are added dropwise. The mixture is stirred over night and filtered off from the precipitated dicyclohexylurea (DCH urea). After evaporation of the solvent, a dark, solid residue remains which is recrystallized from isopropanol. The active ester decomposes on he TLC plate.

Yield 8.65 g (72%); melting point 151° C.
$C_{20}H_{24}N_2O_8$ (420.40)

Calc.:	C 57.17	H 5.75	N 6.66
Found:	C 56.85	H 5.83	N 6.64

H-L-Lys(Bal)-OH

The synthesis is performed according to [17] with 75% yield; melting point 190° C.
$C_{13}H_{18}N_2O_2$ (234.30)

Calc.:	C 64.64	H 7.74	N 11.95
Found:	C 63.44	H 7.35	N 11.03

Z-L-Lys-OH

The synthesis is performed according to [18] with 37% yield; melting point 221°C; $[\alpha]_D^{23} = -10.2 \pm 0.5°$ (c = 2, 0.5 N HCl); uniformly in A and D.
$C_{14}H_{20}N_2O_4$ (280.33)

Calc.:	C 59.98	H 7.19	N 10.00
Found:	C 58.73	H 7.22	N 9.76

Z-L-Lys(Boc)-OH · DCHA

The synthesis is performed according to [18] with 48% yield; melting point 157° C; $[\alpha]_D^{25} = +7.0 \pm 0.5°$ (c = 1, EtOH); uniformly in D.

$C_{31}H_{51}N_3O_6$ (561.76)

Calc.:	C 66.28	H 9.15	N 7.48
Found:	C 66.27	H 9.27	N 7.60

Z-L-Lys(Boc)-OSu 2

14.4 g (37 mmol) Z-L-Lys(Boc)-OH (released with 2 N citric acid from 21.2 g (37 mmol) Z-L-Lys(Boc)-OH · DCHA) and 4.5 g (40 mmol) HOSu are dissolved in 100 ml cold dioxane and 8.25 g (40 mmol) DCCI, dissolved in a little dioxane, are added. After stirring for 16 h, the DCH urea is filtered off, the filtrate evaporated to dryness and the residue crystallyzes from isopropanol.

Yield 14.25 g (81%); melting point 102–103° C; $[\alpha]_D^{23} = -15.1 \pm 0.5°$ (c = 3.5, dioxane).

$C_{23}H_{31}N_3O_8$ (477.51)

Calc.:	C 57.85	H 6.54	N 8.80
Found:	C 57.73	H 6.66	N 8.71

Z-L-Tyr-OMe

The synthesis is performed according to [20] with 79% yield; melting point 94° C; $[\alpha]_D^{23} = -33.0 \pm 0.5°$ (c = 2, DMF); uniformly in D and H.

$C_{18}H_{19}NO_5$ (329.35)

Calc.:	C 65.64	H 5.82	N 4.25
Found:	C 64.74	H 5.90	N 4.22

H-L-Tyr-OMe · HCl

The synthesis is performed according to [19] with 92% yield; melting point 190° C; $[\alpha]_D^{22} = +72.8 \pm 0.5°$ (c = 3.0, pyridine); uniformly in A and D.

$C_{10}H_{14}ClNO_3$ (231.68)

Calc.:	C 51.84	H 6.09	N 6.05	Cl 15.30
Found:	C 51.83	H 6.16	N 6.08	Cl 15.67

Z-L-Tyr-OH

105 g Z-L-Tyr-OMe in 650 ml dioxane/water (4:1) are saponified for 3 h with 305 ml 2 N NaOH. This solution, largely freed from dioxane under vacuum and diluted with water, is acidified with 5 N H_2SO_4 (pH 2) while cooled with ice and stirred. A milky suspension with a brown, oily precipitate is obtained. When the suspension is left to stand, the first fraction Z-L-Tyr-OH crystallizes out in long, white needles from the milky supernatant. The oily residue is taken up in 10% $KHCO_3$ solution at 40° C, hot filtered and the cooled, alkaline filtrate acidified with 5 N H_2SO_4. Under refrigeration, a second fraction crystallizes out; the fractions are dried under high vacuum over P_2O_5.

Yield 34 g (36%); melting point 97–98° C; $[\alpha]_D^{20} = +6.1 \pm 0.5°$ (c = 1, HAc); uniformly in A and D.

$C_{17}H_{17}NO_5$ (315.33)

	Calc.:	C 64.75	H 5.43	N 4.44
Fraction 1	found:	C 64.42	H 5.46	N 4.48
Fraction 2	found:	C 63.19	H 5.41	N 4.43

H-L-Tyr(Bu^t)-OBu^t · HCl 3 · HCl

31 g (0.1 mol) Z-L-Tyr-OH are suspended in a pressure bottle in 800 ml CH$_2$Cl$_2$, the suspension is cooled, and 3.5 ml conc. H$_2$SO$_4$ are added. Quickly 400 ml cooled isobutylene as butylating reagent are added and stirred for 5 days at room temperature (Rt). After the reaction is complete, the reaction mixture is shaken out with cold 10% Na$_2$CO$_3$ solution. The aqueous, alkaline phase is extracted with CH$_2$Cl$_2$, and the pooled, organic phases are filtered, washed neutral with water and dried over Na$_2$SO$_4$. The residue is digested several times with warm hexane after evaporation of the solvent. The pooled hexane phases are filtered and evaporated; an oily residue remains (32.4 g = 75% yield). The oily product Z-L-Tyr(Bu^t)-OBu^t is dissolved in 200 ml absolute EtOH; hydrogenation is performed in 12 h with 1.75 g Pd/C catalyst, adding several drops of glacial acetic acid towards the end of the hydrogenation. The catalyst is filtered over Celite, the filtrate evaporated to approx. 80 ml and shaken with 1 N HCl/CH$_2$Cl$_2$ (1:1). The separated CH$_2$Cl$_2$ phase is shaken once more with 1 N HCl solution, separated and dried over Na$_2$SO$_4$. After evaporation to dryness the residue is taken up in EA and PE added. Recrystallization is done from EA.

Yield 15.1 g (61%); melting point 152° C; [α]$_D^{23}$ = + 39.0 ± 0.5° (c = 3, EtOH); uniformly in A and D.

C$_{17}$H$_{28}$ClNO$_3$ (329.87)
Calc.:	C 61.90	H 8.56	N 4.25	Cl 10.75
Found:	C 61.80	H 8.66	N 4.32	Cl 10.89

Z-L-Asp(β-OBu^t)-L-Val-OH · DCHA 4 · DCHA

3.25 g (28 mmol) L-valine are dissolved in 600 ml 0.05 N NaOH (28 mmol); 11 g (26 mmol) of *1*, dissolved in 300 ml, are added to this solution at 0° C. After 30 h the solution is acidified with 1 N HCl to pH 3 and the suspension extracted with EA. The pooled EA phases are washed with 2 N citric acid and water. After evaporation of the solvent, *4* remains as an oily product, uniformly in systems A and H, which is then converted to the DCHA salt; the crystallization of 4-DCHA is induced by addition of a little PE.

Yield 4.5 g (81%); melting point 122° C; [α]$_D^{20}$ = − 2.4 ± 0.5° (c = 2, EtOH); uniformly in A and H.

C$_{33}$H$_{53}$N$_3$O$_6$ (587.80)
Calc.:	C 65.64	H 9.08	N 6.95
Found:	C 65.84	H 9.03	N 6.49

Z-L-Asp(β-OBu^t)-L-Val-L-Tyr(Bu^t)-OBu^t 5

Dissolve 5.4 g (12.5 mmol) of *4* (released with 2 N citric acid from 8 g (13 mmol) *4 · DCHA*), 4.5 g (13.5 mmol) *3 · HCl*, 1.9 g (14 mmol) HOBt and 1.5 ml N-methylmorpholine in 180 ml DMF and cool to − 12°C. After this temperature is reached, add dropwise 2.87 g (14 mmol) DCCI, dissolved in 10 ml cold DMF. After 36 h, filter off from the precipitated DCH urea, evaporate the solvent under vacuum and take up the oily residue in EA. The EA phase is washed with 2 N citric acid, water and 10% NaHCO$_3$ solution. After washing, filter, dry over Na$_2$SO$_4$, evaporate slightly and induce precipitation with hexane

Yield 1,55 g (18%); melting point 153–155° C; [α]$_D^{22}$ = − 6.0 ± 0.5° C (c = 1, DMF); uniformly in A, D and H.

C$_{38}$H$_{55}$N$_3$O$_9$ (697.87)
Calc.:	C 65.40	H 7.94	N 6.02
Found:	C 65.40	H 8.15	N 6.28

H-L-Asp(β-OBu^t)-L-Val-L-Tyr(Bu^t)-OBu^t · HCl 6 · HCl

Dissolve 1.4 g (2.1 mmol) of *5* in 120 ml MeOH, adding some water, and 150 mg Pd-

(10%)/C-catalyst. Hydrogenate for 60 min in the continuous flow process and evaporate the filtrate several times with MeOH in the rotary evaporator. The residue is taken up in ether and some PE.

Yield 1.17 g (92%); melting point 125–127° C; uniformly in H.
$C_{30}H_{50}N_3ClO_7$ (600.20)

Calc.:	C 60.03	H 8.40	N 7.00	Cl 5.91
Found:	C 59.88	H 8.49	N 7.07	Cl 6.29

Z-L-Lys(Boc)-L-Asp(β-OBu^t)-L-Tyr(Bu^t)-OBu^t 7

Dissolve 950 mg (1.6 mmol) *6 · HCl* in 30 ml DMF and add 0.25 ml (2 mmol) N-methylmorpholine. After 15 min, add 0.78 g (1.6 mmol) of *2*, dissolved in 10 ml DMF, dropwise to the reaction solution. After 48 h, the solvent is evaporated under high vacuum, the residue taken up in EA and filtered off from the insoluble N-methylmorpholine hydrochloride. The EA phase is washed with 2 N citric acid, water, 10% NaHCO$_3$ and water. After drying over Na$_2$SO$_4$ and slight evaporation, 7 crystallizes from ethanol/ice water.

Yield 1.1 g (74%); melting point 144–145° C; $[\alpha]_D^{25} = -13.3 \pm 0.5°$ (c = 0.5, CH$_2$Cl$_2$).
$C_{49}H_{75}N_5O_{12}$ (926.16)

Calc.:	C 63.55	H 8.16	N 7.56
Found:	C 63.00	H 8.23	N 7.71

H-L-Lys(Boc)-L-Asp(β-OBu^t)-L-Val-L-Tyr(Bu^t)-OBu^t · HCl 8 · *HCl*

Dissolve 1.0 g (1.1 mmol) of 7 in 80 ml MeOH, adding a little water, and add 100 mg Pd (10%)/C catalyst. Hydrogenation is performed by addition of methanolic HCl at pH 3.5–4 in the continuous flow process. After 6 h, filter off from the catalyst and evaporate several times with MeOH, take up the residue in ether and wash with water. The organic phase is dried over Na$_2$SO$_4$ and some hexane added.

Yield 650 mg (72%); melting point 165–166° C; uniformly in A, D and H.
$C_{41}H_{70}N_5ClO_{10}$ (828.49)

Calc.:	C 59.44	H 8.52	N 8.45
Found:	C 59.23	H 8.43	N 8.70

Z-L-Arg(Z$_2$)-L-Lys(Boc)-L-Asp(β-OBu^t)-L-Val-L-Tyr(Bu^t)-OBu^t 9

Dissolve 650 mg (0.78 mmol) *8 · HCl* in 30 ml DMF/CH$_2$Cl$_2$ and add 0.11 ml (1 mmol) N-methylmorpholine. After 15 min, add to the solution 450 mg (0.78 mmol) Z-L-Arg(Z$_2$)-OH and 105 mg (0.78 mmol) HOBt. Cool the preparation to −15° C. At this temperature add, dropwise, 165 mg (0.8 mmol) DCCI, dissolved in some cold DMF, to the mixture. After 40 h stirring at Rt, cool the preparation and filter off from the precipitated DCH urea. After the solvent's evaporation and drying under high vacuum, the residue is taken up in 200 ml warm EA. The EA phase is washed with 2 N citric acid, water, 10% Na$_2$CO$_3$ solution and water, filtered, dried over Na$_2$SO$_4$ and ether/PE is added. Recrystallization occurs from DMF/H$_2$O.

Yield 310 mg (30%); melting point 112–114° C; $[\alpha]_D^{22} = -2.2 \pm 0.5°$ (c = 1, DMF); uniformly in A, D and H.
$C_{67}H_{97}N_9O_{17}$ (1295.52)

Calc.:	C 63.01	H 7.45	N 9.32
Found:	C 62.98	H 7.45	N 9.56

H-L-Arg-Lys-L-Asp-L-Val-L-Tyr-OH · 3 HOAc 10 · HOAc

To 310 mg (0.27 mmol) 9 ice-cold and freshly distilled TFA is added and left to stand for 90 min under refrigeration; soon after addition of TFA, complete solution occurs. Then evaporate the solution to approx. half the volume and precipitate the peptide with 200 ml ether. Filter off the precipitate and dry under high vacuum. (Yield 230 mg = 87%.) The intermediate product Z-L-Arg(Z_2)-L-Lys-L-Asp-L-Val-L-Tyr-OH is dissolved in 30 ml MeOH with addition of some water and some DMF, and 30 mg Pd (10%)/C catalyst are added. Deacylation is performed in 2 h by the continuous flow process, then filter off from the catalyst and evaporate the solution. The residue is taken up in some ether and purified on a Sephadex G 15 column (90 × 2.5 cm, eluent: 50 mmol NH$_4$Ac, pH 6.75). The corresponding fractions are collected and evaporated under vacuum to half of their volume, after which lyophilization is repeated.

Yield 200 mg (86%); uniformly in A, J and K.
Amino acid analysis:

	Arg	Lys	Asp	Val	Tyr
Calc.:	1.00	1.00	1.00	1.00	1.00
Found:	0.76	1.08	1.06	1.00	1.00

Racemate test: 1.76% D-Arg, 1.30% D-Lys, 1.50% D-Tyr, 1.66% D-Asp, 0.66% D-Val.

$C_{36}H_{61}N_9O_{15}$ (859.87)

Calc.:	C 50.28	H 7.15	N 14.66
Found:	C 49.94	H 6.88	N 14.46

Animal Experiments

The methylcholanthrene-induced tumor is passaged weekly by intraperitoneal injection of 1×10^6 tumor cells in experimental animals and used for the transplantation 10 days after passage. In addition to TP 5, thymosin β9 (1–7) (Ac-Ala-Asp-Lys-Pro-Asp-Leu-Gly) [21] and a thymus extract [22] (whole fraction) are also tested for their effects on this tumor. The age of the mice is approx. 8 weeks; the experiment is performed in 10 animals per group. After removal from the peritoneum and work-up, the cells are counted under the microscope after staining with trypan blue; <6% dead cells are observed. To establish the tumor, 1×10^5 tumor cells/ml are injected intracutaneously into the abdominal skin. The tumor volume is determined in all 10 animals four times after transplantation at intervals of 7 days; determination is performed by measuring two diameters, a and b, and by calculation according to the formula $V = \frac{2}{3} \pi \, ab \, \frac{a+b}{2}$ · 1 mg peptide resp. peptide fraction in 0.5 ml PBS is injected subcutaneously in each animal at two axillary locations on the first, fourth and seventh day after inoculation.

Summary

A "classical" peptide synthesis for thymopoietin 32–36 (TP5) is described. The effects of TP5, thymosin β9(1–7) and a thymus extract on the growth of fibrosarcoma induced by methylcholanthrene are tested in mice.

Zusammenfassung

Eine „klassische" Peptidsynthese für Thymopoietin 32–36 (TP5) wird beschrieben. Die Wirkungen von TP5, Thymosin β9(1–7) und eines Thymusextrakts auf das Wachstum eines durch Methylcholantren induzierten Fibrosarkoms werden bei Mäusen untersucht.

References

1   Goldstein, A. L.: Recent Prog. Horm. Res. 26: 505 (1970).
2   Trainin, N.: Phys. Rev. 54: 272 (1972).
3   Bach, J. F.; Carnaud, C.: Prog. Allergy 21: 342 (1976).
4   White, A.; Levey, R. N.: Thymic Hormones; in Amos (ed.), Intern. Congr. Immunol., 1st Progress in Immunology, p. 1417 (Academic Press, New York 1971).
5   Pahwa, R.; Ikehera, S.; Pahwa, S.; Good, R.: Thymus 1: 27 (1979).
6   Goldstein, G.: Nature 27: 11 (1974).
7   Schlesinger, D. H.; Goldstein, G.: Cell. 5: 361 (1975).
8   Goldstein, G.; Schlesinger, D. H.; Audhya, T.: Am. Chem. Sci. 20: 6195 (1981).
9   Verhaegen, H.; DeCock, W.; DeCree, J.; Goldstein, G.: Thymus 1: 195 (1980).
10 Goldstein, G.; Schlesinger, D. H.: US-Patent Nr. 4190646.
11 Rampold, G.; Lundanes, E.; Folkers, K.; Voelter, W.; Kalbacher, H.; Bliznakov, E. G.: Z. Naturforsch. 35B: 1476 (1980).
12 Merck: Anfärbereagenzien für Dünnschicht- und Papierchromatographie: 75 (1980).
13 See [12], p. 22.
14 John, W. D.; Young, G. T.: J. Chem. Soc. 662: 2870 (1954).
15 Schröder, E.; Klieger, E.: Liebigs Ann. 673: 208 (1964).
16 See [15], p. 212.
17 Heinzel, W.: Thesis, Tübingen University (1983).
18 See [17], p. 60.
19 Boissonnas, R. A.; Guttmann, St.; Jaquenond, P.-A.; Waller, J.-P.: Helv. Chim. Acta 179: 1491 (1955).
20 Kinoshita, M.; Klostermeyer, H.: Liebigs Ann. Chem. 696: 226 (1966).
21 Müller, J.: Thesis, Tübingen University (1984).
22 Primsch, I.: Diplomarbeit, Tübingen University (1983).

Prof. Dr. W. Voelter Leiter der Abteilung für Physikalische Biochemie des Physiologisch-chemischen Instituts der Universität Tübingen, Hoppe-Seyler-Str. 1, D-7400 Tübingen (FRG)

Antitumoral Action of Xenogenic Substances in Vivo and in Vitro

P. G. Munder[1], Th. Stiefel[2], K. H. Widmann[1], K. E. Theurer[2]

[1] Max Planck Institute of Immunobiology, Freiburg, FRG
[2] Research Laboratories of Organo- and Immunotherapy, Ostfildern, FRG

Introduction

Cell preparations from healthy, non-malignant tissue possess demonstrable tumor-inhibiting properties in various experimental test-systems in vivo and in vitro. Thus, *Wrba* [1] achieved a clear reduction in the tumor-starting rate with methylcholanthrene-induced tumors in mice through prophylactic injections of extracts from bovine decidua. All the animals in the control group had died after 36 weeks, while 40% of the pretreated group were still alive.

Letnansky [2, 3] studied the ^{3}H-thymidine incorporation rate in the DNA in diploid cells and tumor cells under treatment with purified decidual fractions. He found a stimulation of normal cells in this but also a significant inhibition of DNA synthesis in tumor cells. *Paffenholz et al.* [4] confirmed these findings in comprehensive studies of human cell cultures. *Nelson et al.* [5] isolated a DNA-polymerase inhibitor from placenta. *Steinberg et al.* [6] isolated a protein with MG 70 000 from thymus, which likewise has an inhibiting effect on DNA-polymerase. *Ebbesen et al.* [7] demonstrated an up to 80% inhibition of the DNA-synthesis rate of leukemic thymocytes in vitro with thymus and spleen extracts. *Hall et al.* [8] inhibited the growth of leukemia cells with thymus extracts. *Burzynski* [9] isolated a peptide with 15 amino acids taken from the urine of healthy human subjects, which had a significant proliferation-inhibiting effect in tumor-cell cultures. On the other hand, *Thompson et al.* [10] isolated a cytoplasmic factor from tumor cells which stimulated a system-synthesizing DNA in vitro by a factor of 5–10.

These studies, carried out from the aspect of a direct action on tumor cells, are supplemented by the reports of several authors [11–16] who attribute the antitumoral activity of xenogenic proteins to cellular immunological mechanisms. Thus, it was shown in vitro that by incubation of spleen cells with xenogenic serum cytotoxic T-lymphocytes, which can attack and destroy the syngenic tumor cells, are formed in the course of 4–6 days in the culture.

In the present paper, a description is given of prophylactic and therapeutic studies carried out on the Meth-A sarcoma of the mouse with partially sulphated foetal liver tissue. The cytostatic effect was studied on the same tumor model with juvenile, sulphated liver preparations, with non-sulphated liver lyophilisates, with sulphated placenta and thymus preparations, with an xenogenic mixed preparation (NeyTumorin®), and the chemical cytostatic agent cyclophosphamide[1]. In parallel with the animal experiments, the influence of these preparations on the DNA-synthesis rate and the cell count of diploid human fibroblasts and tumor cells in the culture was measured. Here, too, a chemical cytostatic agent (6-mercaptopurine[2]) was used for comparison. Finally, a study was carried out of the effect of partial sulphating on the electrophoretic pattern of liver preparations.

Material and Methods

Sample Preparation

Experiments with partially sulphated organs were carried out with preparations from the vitOrgan Company, Ostfildern, FRG. For the animal experiments, all preparations used were centrifuged at 100 000 g for 2 h, sterilized by ultrafiltration and stored at −20° C. The NeyTumorin® (vitOrgan) used as the xenogenic mixed preparation, consists of a mixture of bovine and porcine thymus juv., placenta mat., liver, and other organs such as the pancreas.

For the sulphating, the lyophilized tissue is exposed for 40 h to the gas phase under concentrated sulphuric acid at reduced pressure (10^{-2} bars). The dry substance thus obtained was suspended in citrate phosphate buffer (pH 7.2), homogenized for 10 min in the Ultraturrax (Janke und Kunkel, Staufen i. Br.) and centrifuged. The supernatant was sterilized by filtration and the protein content determined by *Lowry*'s method [17].

[1] cyclophosphamide: Endoxan® (Asta-Werke, Bielefeld, FRG)
[2] 6-mercaptopurine: Sigma (Munich, FRG)

Polyacrylamide Electrophoresis (PAG E)

The PAG electrophoresis was carried out with 7.5% gels (T = 7.7%, C = 2.6%) in the tris-glycin buffer system (pH 8.9). The 10 µl aliquotes of the sample solution with a protein content of 1–10 mg/ml was pipetted into poured trays and separated at initially 10 mA/220 V for 15 min and then 40 mA/300 V for 4 h. After fixing, staining with Coomassie brilliant blue 250R and exhaustive decolorization was carried out.

Animal Experimentation

For the tumor studies in vivo, (Balb/c × C57b16)Fl mice were used (Jackson Lab., Bar Harbor/Me., USA). The Meth-A Sarcoma [18] was passaged weekly in these animals by i.p. injections of 1×10^6 tumor cells and used for transplantation 7–10 days after the passage.

For this, 1×10^5 tumor cells were injected i.c. in the abdominal skin. After the exponential growth of the tumor, the untreated animals died after 4–5 weeks. The studies were carried out with 10 animals per group. Treatment of the verum groups was carried out by one to three i.m. or i.v. injections of 1 mg each (in relation to protein) of the organ preparations.

The tumor volume (TV) was established by measurement of two diameters using by the formula $V = \frac{2}{3} \pi ab \frac{a+b}{2}$. The TV quoted is the total volume per group.

Cell-Culture Technique

Human tumor cells (melanoma, Wish, Seromed, Munich, FRG) and diploid fibroblasts (F.H.2: biopsy cells from healthy child; MRC-5: Seromed, Munich) were used as test cells. All cell cultures passed through a test-cycle consisting of two growth-phases and a minimal phase. The test-substances were added in the minimal phase.

Growth Phase 1

The cells, which were kept in liquid nitrogen, were built up at 37° C and incubated in culture flasks, together with MEM-H medium (Instamed, Seromed, Munich) with additions of penicillin, streptomycin, neomycin, and 10% foetal-calf serum (FCS). After 2 days, the cells were visually counted after trypsinization and seeded again (2×10^6 cells per flask).

Growth Phase 2

After a daily change of the medium, the cell-count was again carried out on the 4th day and 0.5×10^6 cells/10 ml MEM + 10% FCS seeded in each 30 ml culture flask.

Minimal Phase

After 24 h, the medium was drawn-off and a minimal medium with reduced serum admixture (diploid cells: 1%; heteroploid cells: 0.25%) added.

Test Phase

After 48 h, the preparation was added, 50 µl preparation being used per flask in the present tests. Three parallel samples were used per test for each kind of preparation and each concentration. For the short-period tests, incubation was limited to 8 h. 4 h before the end of this exposure period, 50 µl methyl-^{3}H-thymidine solution (0.35 µCi) was added by pipette in each case. After 8 h, the non-incorporated thymidine was removed by washing four times with 5 ml 2% perchloric acid, in each case, and hydrolizing the cells with 1 N HCl at 70° C. The hydrolysate was transferred to vials and mixed with scintillation cocktail (Aquasol 2, NEN). The activity measurement was carried out in a Beckman LS-100 Liquid Scintillation Counter. Relative standard deviations of the total test system of $s_r = \pm 10\%$ were determined as statistical quality coefficients.

Long-Term Studies

The long-term experiments were largely carried out in agreement with the short-period tests. However, it was not possible to determine any ^{3}H-thymidine incorporation rates; the cell count was determined by visual means.

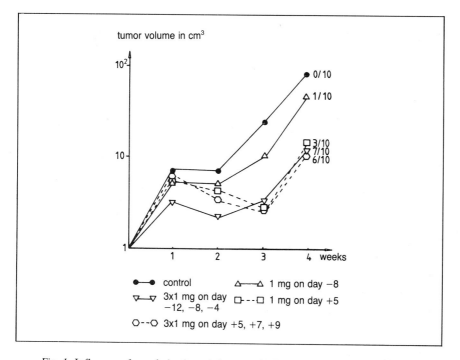

Fig. 1. Influence of prophylactic and therapeutic doses of sulphated foetal liver tissue on the Meth-A sarcoma in vivo. X/10 = survivors/total number per group. Tumor volume is the total volume of all tumors in the group.

Results

The first animal studies with the methyl-cholanthrene-induced fibrosarcoma (Meth-A sarcoma) in the mouse were carried out as a prophylactic measure. For this, each animal was either treated only once on the 8th day or three times on the 12th, 8th, and 4th days with 1 mg foetal liver preparation in each case, before tumor cell injection. After the tumor implantation, the tumor volume was measured weekly and the survival rate determined after 4 weeks. It was found (fig. 1) that even a single prophylactic treatment clearly reduced the tumor volume after 4 weeks. However, the survival rate was only 1 of 10 animals, compared with none in the control group. The triple prophylactic treatment raised the survival rate to 70% and reduced the tumor volume to about 20% of the untreated control.

The therapeutic tests on the same tumor system confirmed the tumor-inhibiting effect. Foetal liver tissue proved to be especially potent when injected three times i.m. or i.v. on the +5th, +7th, and +9th day after the transplantation of the tumor. Here, too, the triple treatment was shown to be more effective than a single injection, as demonstrated above all by the number of surviving animals. The course of the curve also shows, that

(1) the tumor growth is the same in all groups during the 1st week;

(2) while the tumor growth in the 2nd week is somewhat slowed down in the control group, the tumor in both the prophylactically and therapeutically treated groups regresses in the 2nd to the 4th week;

(3) that after the end of the stagnation or regression phase, the tumor again grows normally in an exponential manner in animals which have not been cured.

To obtain information about possible cross-reactions between tumor-specific transplantation antigens and oncofoetal antigens in the xenogenic preparations, a comparison was made of the efficacy of the foetal and juvenile liver preparations. In several tests, it was shown that both preparations were characterized by the same potency. Both the tumor volume and the survival rate were identical, as shown by figure 2, for example. The purpose of further studies with this tumor model was to determine whether partially sulphated preparations differ from pure lyophilisates in their antitumoral action.

There is no qualitative difference as shown by figure 3, both juvenile lyophilized liver tissue and additionally sulphated liver tissue inhibit tumor

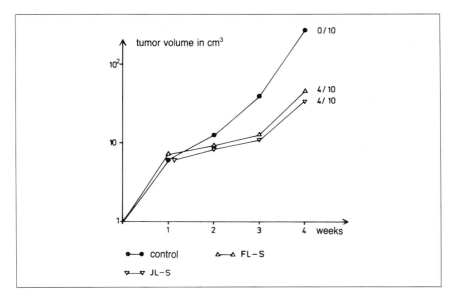

Fig. 2. Antitumoral effect of foetal (FL-S) and juvenile (JL-S) liver tissue on the growth of Meth-A sarcoma cells in vivo. For tumor volume, see figure 1.

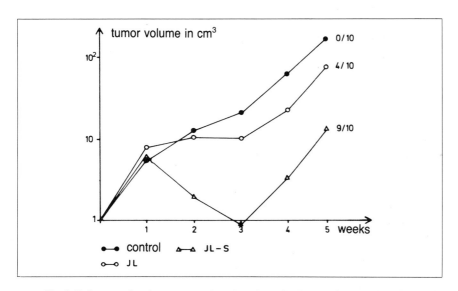

Fig. 3. Influence of various preparations from juvenile liver on the growth of the Meth-A sarcoma in the mouse (JL: juvenile liver lyophilisate; JL-S: partially sulphated juvenile liver tissue).

growth and increase the survival rate. There was, nevertheless, a quantitative difference: While the pure lyophilisate in this test displayed a survival rate of 40%, the treatment with sulphated tissue gave a survival rate of 90%. An analytical examination of the lyophilisate and sulphated liver preparation from identical organs, which was then carried out, revealed a modified mobility of liver proteins after sulphating in the electrophoretic pattern. Figure 4 shows a schematic representation of the PAG electrophoreses of lyophilized and sulphated foetal and juvenile liver preparations. It will be noted that with a comparable protein spectrum, the sulphated samples display an increased electrophoretic mobility.

The antitumoral action is not liver-specific. Lyophilisates or additionally sulphated preparations from other xenogenic tissues also possessed an antitumoral effect in this system (fig. 5). The most potent antitumoral effect was displayed by mixtures (1:1) of tissues of different species such as cattle and pigs (NeyTumorin®). These lyophilisates can bring about a complete regression in up to 100% of the growing tumors, as shown by figure 5. Finally, the antitumoral effect of the liver preparations was also compared with that of a classic chemocytostatic agent in the Meth-A system (fig. 6). For this, 1 mg liver preparation or 2 mg cyclophosphamide was injected on the + 4th, + 6th and + 8th day. After treatment with the chemocytostatic agent, only 1 animal of 10 survived the 4-week period of observation, whereas 8 of the 10 animals in the group treated with the liver preparation survived. There is a striking difference in the development of the tumor volume. As expected, cyclophosphamide has an immediate cytostatic effect which disappears after the 1st week when discontinued. In contrast, as mentioned, the liver preparation does not

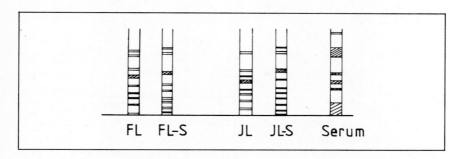

Fig. 4. Electrophoretic patterns of various liver preparations in comparison with standardized human serum (FL: foetal liver lyophilisate; FL-S: partially sulphated foetal liver tissue; JL: juvenile liver lyophilisate; JL-S: partially sulphated juvenile liver tissue).

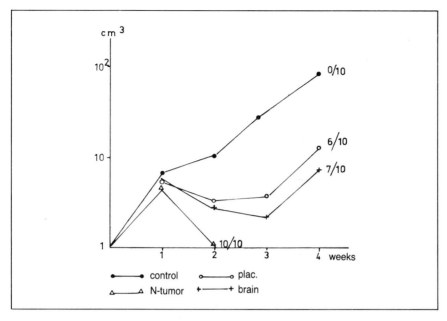

Fig. 5. Comparison of the antitumoral effect of various tissues (placenta, brain) and a mixture of various xenogenic tissues (NeyTumorin®).

initially have any influence on the progression of the tumor but in the 2nd and 3rd weeks induces an increasing tumor regression. Side effects were not observed in any of the tests. No anaphylactic reactions whatsoever were observed, even with the 10 × i.v. application of 1 mg liver preparation for a period of 6 months – and more – in mice and rats.

At the same time, human cells (melanoma and diploid fibroblasts) were studied with respect to their synthesis and division behavior in the presence of preparations successfully tested in animals. The cytostatic effects of the preparations were compared with that of the chemocytostatic agent 6-mercaptopurine (6-MCP). The daily incubation of 2×10^6 cells with 5×10^{-7} g liver preparation led, in tumor-cell cultures (Wish) after 20 days, to a decline in the cell count to max. 60% of the control (± 13.5; p 0.05), whereas normal cells (diploid fibroblasts F.H.2) are not inhibited but slightly stimulated, even if not to a statistically significant extent. With the chemocytostatically treated cultures, a clearly, more potent cytostatic effect was observed in the tumor-cell culture. The nor-

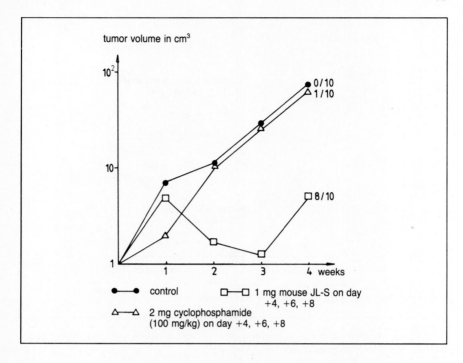

Fig. 6. Comparison of the tumor-inhibiting effect of cyclophosphamide and liver tissue on Meth-A sarcoma cells in the mouse (JL-S: partially sulphated juvenile liver tissue).

mal-cell culture was likewise damaged and reacted, after 20 days, with a decline in the cell-count to about 40% of the control (fig. 7).

When sulphated thymus preparation was used, a dose-related inhibition of the DNA synthesis of melanoma cells to a max. 45% ($\pm$ 15%) of the untreated controls was found after already 8-h incubation. The most marked effect was at a dose of 2×10^{-5} g protein/0.5×10^6 cells. However, with the same dose, diploid fibroblasts could be stimulated to a DNA-synthesis rate of almost 150% ($\pm$ 16%) in comparison with the control (fig. 8).

Finally, a study of the dose-effect relation of NeyTumorin® in melanoma cell cultures was carried out. An almost linear rise in the DNA-synthesis inhibition was found in the range 0.03–0.1 mg protein/10^6 cells (fig. 9). The data are mean values taken from 13 tests with a mean standard deviation per point of 15.5%.

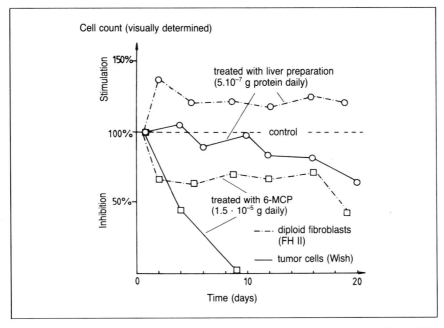

Fig. 7. Course of cell counts of human cell cultures after treatment with partially sulphated liver preparation and 6-mercaptopurine (mean of 2 tests, 3 × determinations).

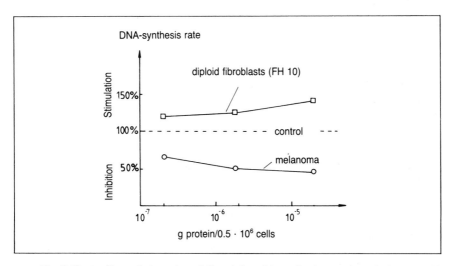

Fig. 8. Dose-effect relation of partially sulphated juvenile thymus tissue in human cell culture.

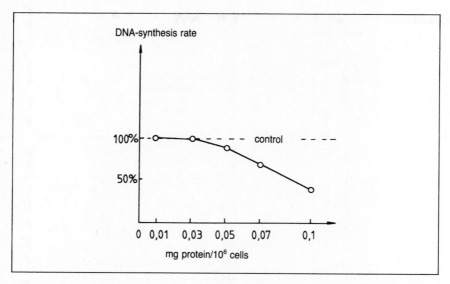

Fig. 9. Dose-effect relation of an xenogenic organic mixture (NeyTumorin®) in the melanoma cell culture.

Discussion

The present studies show a prophylactic and therapeutic effect of xenogenic tissue preparations in the Meth-A sarcoma model of the mouse. A tumor-inhibiting action was also demonstrated on human tumor cells in vitro, but the growth of normal cells was not impaired by the same preparations. A dose-dependence of the effect of these preparations was found in animal experimentation, inasmuch as concentrations of less than 0.1 mg/animal were ineffective; 0.5–1 mg were shown to be optimal with a triple application. On the other hand, with the cell-culture tests, a clearly stricter dose-effect relation was found (fig. 9). If this tumor-inhibiting effect is compared with that of classic chemocytostatic agents, it was shown in the animal study, that under given test-conditions the cyclophosphamide effect could be significantly surpassed by administrating the liver preparation in respect to tumor growth and survival rate. In the human cell culture, the preparations studied selectively inhibit the tumor cells, while 6-mercaptopurine, which was studied simultaneously, had an inhibiting action on both tumor and normal cells.

Since foetal liver tissue was successfully used in the first tests, it appeared likely that there is a structural relation between a tumor-specific transplantation antigen of the Meth-A sarcoma and an oncofoetal antigen. This leads to a humoral or cellular immunological response when foetal liver tissue is administered. The ensuing studies with juvenile liver preparations showed, however, that neither the tumor growth nor the survival rate were influenced differently by foetal or juvenile preparations (fig. 2). It thus appears that the mechanism of action is not based on a cross-immunization by oncofoetal antigens.

In comparative studies on Meth-A sarcoma, an inhibition of tumor growth was observed, but with quantitative differences for both pure liver lyophilisates and for sulphated liver tissue. Whether the increased hydrophilia of the compound molecules, due to the sulphating, can be made responsible for this enhancement of the cytostatic effect must remain an open question for the time being. Paralleling this, there is an increased electrophoretic mobility, presumably due to the addition of SO_3-H groups.

The antitumoral effect is not restricted to liver tissue. Both brain and placenta preparations displayed tumor-inhibiting properties without, as a rule, being able to attain the values of liver preparations. The most potent cytostatic effect was attained by a xenogenic mixed preparation (Ney-Tumorin®). This mixed preparation was usually capable of bringing about a complete regression of a normally growing Meth-A sarcoma. An endotoxin effect on the Meth-A sarcoma could be excluded, since the preparations used in the pyrogenicity test in the rabbit were free of endotoxin. Finally, fairly recent studies show that the antitumoral action of the xenogenic material does not depend on the tumor. Both the 3-Lewis lung carcinoma in C57bl/6 mice and the L1210-suspension tumor in DBA mice are markedly inhibited (in preparation).

From the available data, several mechanisms can be discussed. Our own unpublished studies of a possible activation of bone-marrow macrophages show that these cells, after incubation with xenogenic, foetal and also juvenile liver tissues, can destroy tumor cells in vitro with increased capacity.

On the other hand, studies of a series of authors show that foetal calf serum, for example, can induce cytotoxic T-lymphocytes in the mouse in vitro, which can also destroy syngenic tumor cells [11–16]. Our own studies in vitro with spleen cells of the mouse and various murine and also human tumor cells as target cells, confirm these findings (in preparation).

They show that after 6 days of incubation with xenogenic liver preparations, one and the same spleen-cell population can destroy several different syngenic and allogenic tumor cells within 24–48 h. A polyclonal stimulation of the spleen-cell population by xenogenic determinants would appear possible [16]. These in vitro findings thus also permit a rational interpretation of the present in vivo tests. This assumption is supported by the observation, that also in vivo the xenogenic preparations only take effect after about a week, i.e., after a necessary time for the development of an immunological reaction. Furthermore, the antitumoral effect of xenogenic material could not be demonstrated in the present tests in thymusless nu/nu mice, nor in animals exposed to 450 rad.

However, the results presented from the cell-culture also indicate that there are factors in the preparations, which selectively inhibit the DNA-synthesis of tumor cells. Consequently, a direct effect also appears possible, in addition to the cellular immunological reaction mentioned.

Whether this is true or not, further clarification of the mechanism of action presupposes the separation of the complex organic lysates. Initial studies suggest that when customary chromatographic and electrophoretic methods are used biologically active compounds are to be sought in the < 10 000-MG range.

The efficacy with which the growth of tumors in vivo and in vitro is inhibited by the application of the xenogenic tissue preparations described, makes further analysis of this unexpected antitumoral mechanism of action a compelling necessity.

Summary

The proliferation of various tumor cells was inhibited in vivo and in vitro after application of or incubation with xenogenic liver tissue. The development of s.c.-implanted Meth-A sarcoma was blocked by the prophylactic injection of these preparations. In addition, firmly established tumors regressed under therapy. Preparations obtained from xenogenic organs such as the thymus, placenta or brain, had a similar antitumor activity. A mixture of various xenogenic tissues from different species had a much higher therapeutic efficiency. In the Meth-A system, the xenogenic material surpassed the antineoplastic effect of just-tolerable doses of cyclophosphamide. The preparations showed no side effects in mice and rats. These results were supported by experiments in tissue culture on human tumor cells. This new antitumoral activity of xenogenic tissues in vivo is interpreted as a stimulation of the body's own resistance. The results in tissue culture, however, are also an indication of a direct regulatory effect on cells.

Zusammenfassung

Die Proliferation von Tumorzellen wurde durch Applikation oder Inkubation mit xenogenen Leberpräparationen in vivo und in vitro gehemmt. Prophylaktische Injektionen dieser Präparate hemmten die Entwicklung subkutaner implantierter Meth-A-Sarkome. Darüber hinaus konnten etablierte Tumoren durch Therapie zur Regression gebracht werden. Präparationen aus anderen xenogenen Organen wie Thymus, Plazenta und Hirn hatten eine ähnliche Wirksamkeit. Besonders effektiv war eine Mischung aus verschiedenen xenogenen Materialien. Die antitumorale Wirkung von noch tolerierbaren Dosen von Cyclophosphamid konnte im untersuchten Meth-A-Sarkom übertroffen werden. Diese Befunde konnten durch Ergebnisse in der Gewebekultur an humanen Tumorzellen gestützt und erweitert werden. Diese neuartige, antitumorale Wirkung xenogener Gewebe in vivo wird als eine Verstärkung der körpereigenen Abwehr interpretiert. Die Ergebnisse der Gewebekultur sprechen aber auch für einen direkten regulativen Effekt auf die Zelle.

References

1. Wrba, H.: Krebsverhütung und Verhinderung der Krebsentstehung. Österr. Ärztez. 29: 1351–1352 (1974).
2. Letnansky, K.: Stoffwechselregulatoren der Plazenta und ihre Wirkung in Normal- und Tumorzellen. Exp. Path. 8: 205–212 (1973).
3. Letnansky, K.: Tumorspezifische Faktoren der Plazenta und Zellproliferation. Exp. Path. 9: 354–360 (1974).
4. Paffenholz, V.; Theurer, K.: Einfluß von makromolekularen Organsubstanzen auf menschliche Zellen in vitro. I. Diploide Kulturen: Z. Kassenarzt 27: 5218–5226 (1978); II. Tumorzellkulturen: Z. Kassenarzt 19: 1876–1887 (1979).
5. Nelson, J. A.; Levy, J. A.; Leong, J. C.: Human placentas contain a specific inhibitor of RNA-directed DNA-polymerase: Proc. natn. Acad. Sci. USA 78: 1670–1674 (1981).
6. Steinberg, J. A.; Otten, M.; Grindey, G. B.: Isolation of DNA Polymerase-associated Regulatory Protein from Calf Thymus. Cancer Res. 39: 4330–4335 (1979).
7. Ebbesen, P.; Olsson, L.: Stimulatory Effect on DNA-Synthesis of Thymus and Spleen Extract from Leukemic AKR Mice. J. Cancer Res. Clin. Oncol. 100: 105–107 (1981).
8. Hall, V.; Wolcott, M.: Modulation of Tumor Cell Growth by Thymus Extracts. Fed. Proc. 40: 3351 (1981).
9. Burzynski, S. R.; Stolzmann, Z.; Szopa, B.; Stolzmann, E.; Kaltenberg, O. P.: Antineoplaston A in Cancer Therapy. Physiol. Chem. Phys. 9: 485–500 (1977).
10. Thompson, L. R.; McCarthy, B. J.: The effects of cytoplasmatic extracts on DNA-synthesis in vitro. Biochim. Biophys. Acta 331: 202–213 (1973).
11. Tsutsui, J.; Everett, N. B.: Specific versus nonspecific target cell destruction by T lymphocytes sensitized in vitro. Cell. Immunol. 10: 359–370 (1974).
12. Levy, R. B.; Shearer, G. M.; Kim, K. J.; Asofky, R. M.: Xenogenic-serum-induced murine cytotoxic cells. Cell. Immunol. 48: 276–287 (1979).
13. Kedar, E.; Schwartzbach, M.: Further characterization of suppressor lymphocytes induced by fetal calf serum in murine lymphoid cell cultures: comparison with in vitro generated cytotoxic lymphocytes. Cell. Immunol. 43: 326–346 (1979).
14. Watson, J.; Gillis, S.; Marbrook, J.; Mochizuki, D.; Smith, K. A.: Biochemical and biological characterization of lymphocyte regulatory molecules. J. exp. Med. 150: 849–861 (1979).

15 Golstein, P.; Luciani, M. F.; Wagner, H.; Röllinghof, M.: Mouse T cell-mediated cytolysis specifically triggered by cytophilic xenogenic serum determinants: a caveat for the interpretation of experiments done under "syngenic" conditions. J. Immun. *121:* 2533–2538 (1978).
16 Golstein, P.; Rubin, B.; Denizot, F.; Luciani, M. F.: Xenoserum-induced Cytolytic "T" Cells: Polyclonal Specificity with an Apparent "Anti-Self" Component and Cooperative Induction. Immunol. biol. *156:* 121–137 (1979).
17 Lowry, O. H.; Rosebrough, N. J.; Farr, A. L.; Randall, R. J.: Protein Measurement with the Folin Phenol Reagent. J. biol. Chem. *193:* 265–275 (1951).
18 Lloyd, L. J.; Boyse, E. A.; Clarke, D. A.; Carsweill, E.: The antigenic properties of chemically-induced tumors. Ann. N.Y. Acad. Sci. *101:* 80–106 (1962).

Dr. P. G. Munder, Max-Planck-Institut für Immunbiologie, Stübeweg 51, D-7800 Freiburg (FRG)

The Inhibition of Tumor Proliferation by Specific Factors Separated from the Maternal Part of the Bovine Placenta

K. Letnansky

Institute of Cancer Research, University of Vienna (Austria)

Cell proliferation is controlled at different levels in a very complex manner. One of the factors involved in this process is the influence of biological components which are capable of stimulating or inhibiting DNA-synthesis. Some of these regulators, including the growth factors [7, 21], have been characterized in recent years. Most of them are bound to the cell surface in a specific manner – a process which is followed by the rapid internalization of the receptor-ligand complex.

Among the family of growth inhibitors, the chalons have been investigated most extensively (reviewed by [22]). One of the most typical properties of these compounds is their tissue specificity; a property which differentiates them from a series of other inhibitors of DNA-replication. Another type of specific action was discovered in preparations made from bovine placentas, namely an inhibition of DNA-synthesis and cell proliferation, directed predominately against tumor cells [13]. Investigations into the nature of this factor as well as into the mechanism of its action are described in this paper. According to the results here presented, binding the inhibitor to surface receptors appears to play a critical role in the development of cellular specificity. In this respect, the mechanisms ultimately resulting in the inhibition of DNA-replication apparently show some similarity to those involved in the stimulating action of growth factors or insulin.

The varying effects of placental preparations, obtained by a lyophilization process as described by *Theurer* and *Triebel* [14], on the in-vitro incorporation of inorganic phosphate or amino acids into normal and tumor cells, have been described in a series of earlier experiments. In addition, differences concerning the degree and mode of action of factors in preparations obtained from the fetal or maternal part of bovine placentas have been reported [6, 9, 26].

Most interestingly, in-vivo experiments demonstrated that mice-injected preparations from deciduas resulted in a significantly diminished tumor induction by 3-methylcholanthrene [27]. Another model, the viviparous platyfish Platypoecilus maculatus, known for its high rate of spontaneous melanoma formation, demonstrated a significantly diminished tumor incidence as a consequence of the treatment with extracts from maternal placentas during the embryonal phase [8].

All these experiments were carried out with the unfractionated preparations, obtained by the lyophilization procedure. However, it turned

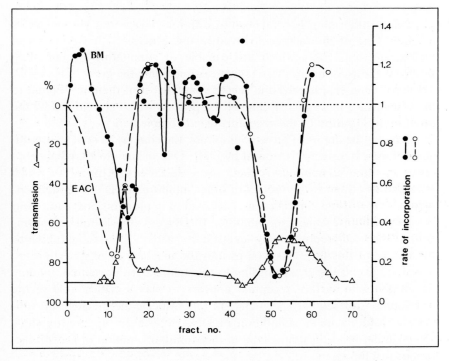

Fig. 1. Fractionation of bovine maternal placenta on a Sephadex G-100 column by elution with TKM buffer. Transmission of the eluting material at 260 nm (△—△) and inhibition of thymidine incorporation into the DNA of bone marrow (●—●) and Ehrlich-ascites tumor cells (o----o) are shown for individual fractions.

out that the maternal portion of the placenta contained not only three different substances with the capacity to inhibit cellular DNA-synthesis, but also some other components, stimulating the incorporation of thymidine into high-molecular weight material.

Separation of the original material by means of molecular-weight exclusion chromatography with Sephadex G-100, resulted in the isolation of two inhibitors with very similar molecular weights. However, they could be distinguished by their clear-cut specificity regarding tumor and normal cells [13]. A third component was of lower molecular weight and showed no specificity (fig. 1). Further purification of the material displaying tumor-specificity was achieved by ion-exchange chromatography with Dowex 50W (manuscript in preparation).

The apparent molecular weight of the inhibitor, as determined by SDS-polyacrylamide gel electrophoresis, is higher than 60,000. The activity of the factor is significantly reduced by its incubation with proteolytic enzymes, including trypsin and papain. It may be concluded from this that DIF («decidua inhibitory factor») is a protein or a glycoprotein [15].

In-vitro experiments demonstrated the specific action of DIF on a variety of tumor cells. The incorporation of ^{3}H-thymidine into high-molecular weight material of the transplantable Yoshida-ascites sarcoma, Ehrlich-ascites carcinoma, lymphatic leukemia cells as well as the cells of established lines of a human osteosarcoma (line 2T) and a human bronchogenic carcinoma (squamous cell carcinoma; line E14) is inhibited to values between 16% and 59% of the controls. On the other hand, thymidine incorporation into normal cells, including rat bone marrow and the fibroblast cell line Wi38, was inhibited to values not less than approximately 95% of the controls [18].

The inhibition of DNA-synthesis obviously results in a severe damage of the tumor cells, since tumor development and growth is significantly reduced after the inoculation of cells preincubated with DIF for 2 h at 37°C [14] (fig. 2).

This effect of DIF on DNA-synthesis is not observed when cell-free preparations are used as a test system (i. e., nuclear fractions or nuclear extracts plus activated external DNA, according to the methods of *Thompson et al.* [25], and *Burke et al.* [1]). This could be interpreted that the DNA polymerase reaction itself is not inhibited, but rather one of the steps before this event.

One of the possible causes of this might be an ATP deficiency, resulting from an inhibited substrate utilization. However, ATP determinations

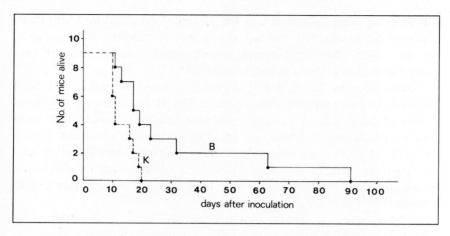

Fig. 2. Survival of mice, inoculated with Ehrlich-ascites tumor cells, which had been incubated in the presence (B) or absence (K) of DIF. Incubation was performed with 2 ml of a cell suspension, prepared as described under «Methods», and 1 ml of inhibitor-fraction obtained as described in figure 1 (B), or 1 ml TKM buffer (K). Duration of incubation was 2 h at 37°C.

in tumor cells incubated in the presence of DIF showed no influence on the ATP level, as compared to the uninhibited control cells. Moreover, phosphorylation processes seem to proceed in a normal way, as demonstrated by polyacrylamide gel electrophoresis of nuclear phosphoproteins [15].

The latter results also indicate that major alterations in chromatin conformation, which could result in an altered genetic activity and are reflected in the phosphorylation pattern of nuclear proteins ([16, 17, 19], and literature cited there) are not induced by the inhibitor.

According to these and other results, the involvement of membrane phenomena in the course of the inhibition of tumor DNA replication seems very likely. This is also indicated by the observation that the inhibitory action of DIF is dependent on its concentration: increasing amounts of DIF result in an increasing inhibition of thymidine incorporation into Ehrlich tumor cell-DNA in a manner characteristic to saturation kinetics (fig. 3). This indicates the existence of specific receptors on the plasma membranes of the cells used in these experiments.

Therefore, binding-studies were performed with the radioactively labeled inhibitor to compare membrane preparations obtained from normal and from tumor cells. These investigations clearly demonstrate that

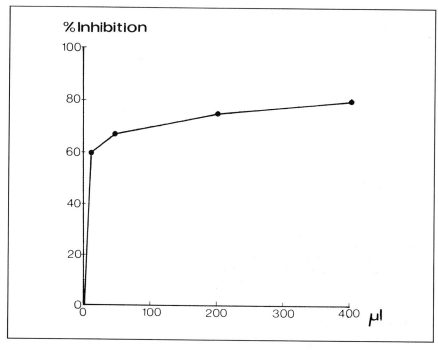

Fig. 3. Inhibition of ^{3}H-thymidine incorporation into the DNA of Ehrlich-ascites tumor cells dependent on the amount of DIF containing column effluents.

significantly higher amounts of DIF are bound to the tumor membranes than to membranes of the normal control cells: while Yoshida-tumor membranes bound DIF in an amount corresponding to 5,500 dpm/µg of membrane protein, only 3,500 dpm/µg were bound to rat bone-marrow membranes.

Since this could be the result of changes in binding affinities and capacities as well, the dependence of the binding reaction on DIF-concentrations was compared with tumor and liver cell membranes and evaluated by the use of *Scatchard* plots [23]. As shown in figure 4, surface membranes from normal cells, such as liver, do not bind DIF as efficiently as do tumor membranes. In Yoshida membranes, the binding capacity is about twice as high as in normal surface membranes, but no significant difference is observed concerning the affinities of the receptors for DIF. However, an additional class of receptors can be detected on tumor sur-

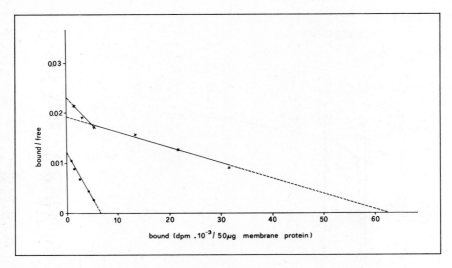

Fig. 4. Scatchard plot [23] of rat liver membranes (•——•) and Yoshida-tumor cell membranes (x——x) binding ³H-propionylated DIF as described under «Methods».

faces, which results in a ten times higher binding-capacity (even though they demonstrate a lower affinity for the inhibitor).

Taken together, these results favor the hypothesis that the in vitro action of the inhibitor is primarily dependent on its binding to specific cell surface receptors and on the subsequent internalization of the resulting receptor-ligand complex. The results further indicate the existence of additional receptors on tumor plasma membranes, resulting in a more efficient internalization of the inhibitor by these cells.

Although it cannot be excluded that these additional receptors exist also in a cryptic form on normal cells, the following experiments rather indicate their specific existence on tumor cells. In one of these experiments, surface membranes from rat liver and tumor cells were separated by SDS-polyacrylamide gel electrophoresis and their composition was compared after protein and carbohydrate staining. As demonstrated in figure 5, there are indeed significant differences between membrane components of liver cells compared to those of a hepatoma or the Yoshida sarcoma. This not only applies to the quantitative and qualitative distribution, but also to the carbohydrate content of some components.

To elucidate which of these compounds is responsible for the enhanced and/or additional binding of the inhibitor, membrane prepara-

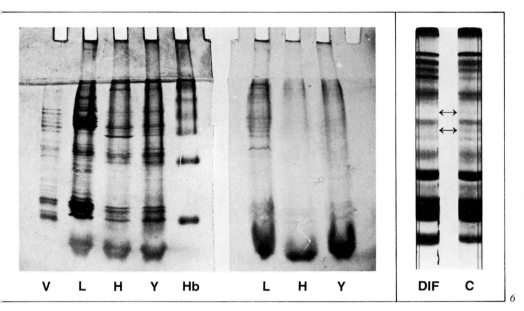

Fig. 5. Separation of plasma membranes from rat liver (L), Yoshida-ascites tumor (Y), and diethylnitrosamine-induced hepatoma (H) on 15% SDS-polyacrylamide gels, stained with Coomassie blue for proteins (left gel) and with PAS (periodic acid, Schiff reagent) for carbohydrate (right gel). Hb: molecular weight standard «cross-linked hemoglobin» with bands corresponding to M_r = 16,000, 32,000, 48,000, 64,000, 80,000, and 96,000. V: Nuclear proteins from Yoshida cells.

Fig. 6. Separation of nuclear proteins from Yoshida cells which had been incubated for 1 h at 37°C in the presence (DIF) and in the absence (C) of DIF on 15% acid-urea polyacrylamide gels. Gels are stained with amido black.

tions from liver and Yoshida-tumor cells were separated on acid-urea gels and sliced into 1 mm disks. These were then incubated with the tritium-labeled inhibitor. It turned out that enhanced binding of DIF took place primarily to receptors in the molecular weight range between 20,000 and 40,000, and to one receptor with about 60,000 daltons (manuscript in preparation). This again agrees with the presence of higher intracellular DIF concentrations in tumor cells and a more effective inhibition of one of the reactions in the complex sequence ultimately resulting in DNA-replication.

According to recent experiments from Pardee's laboratory [2, 10, 20], a specific «R-protein» is one essential component of the complex DNA polymerase system. Molecular weights in the region of about 53,000 daltons have been reported ([2–4, 10, 20], and literature cited there), depending on the source of the R-protein. This protein is very labile and a constant level of this essential compound is maintained only in cells with normally operating protein synthesis. Inhibition of protein synthesis, for instance by cycloheximide, results in a rapid cell-loss of R-protein.

According to earlier results from this laboratory, DIF not only influences the synthesis of DNA but also that of proteins. Although apparently non-specific, a significant inhibition of amino-acid incorporation into high-molecular weight material has been observed, using cell-free preparations of rat liver and Ehrlich-ascites carcinoma cells. In these experiments, it was further shown that the extent of the inhibition is proportional to the concentration of the inhibitor [11].

The demonstration of a decrease in proteins with a molecular weight of about 50,000 to 60,000 under the influence of DIF would strongly support an explanation of inhibitory action involving the R-protein. We, therefore, compared the patterns of nuclear proteins from Yoshida-ascites sarcoma cells, which were incubated in the presence and absence of DIF. As shown in figure 6, incubation with DIF actually results in a significant decrease in two protein fractions, the molecular weights of which were about 48,000 and 65,000 daltons.

To summarize, the experiments presented here describe the separation of a factor from the maternal part of bovine placentas, capable of inhibiting protein and DNA synthesis in a specific manner. As a result, this factor inhibits the proliferation of tumor cells. The experiments further document an enhanced binding (and subsequent internalization) of this factor to receptors of the tumor plasma membrane. This process presumably results in an increased inhibition of the synthesis of a protein, essential in the assembly of the DNA-polymerase multienzyme complex.

Methods

Separation of Decidua Extracts on Sephadex G-100

Extracts from the maternal part of the bovine placenta were prepared either by homogenization of fresh material in 3 volumes of TKM buffer (Tris/HCl pH 7.4, 10^{-2}M, KCl, 10^{-2}M, MgCl$_2$ 1.5 × 10^{-3}M) and subsequent centrifugation or by dissolving a lyophilized preparation, obtained from vitOrgan, Stuttgart, in the same buffer. Extracts were then separated on Sephadex G-100 columns with a length of 45 or 85 cm and being 1 or 2.6 cm in diameter by elution with TKM.

Incorporation of Thymidine into High-Molecular Weight Material

Cells were suspended in 8 ml RPMI 1640, containing 84 mg morpholinopropanesulfonic acid (in earlier experiments 1 ml RPMI was replaced by 0.7 ml fetal-calf serum and 0.3 ml horse serum). Then, to 400 µl of this suspension was added: 100 µl (corresponding to 2 µCi) of ^{3}H-thymidine (spec. activity 1 mCi/ml), diluted with the above mentioned medium, and 100 µl of individual fractions obtained by the chromatographic separation procedure. After incubation at 37°C for 20, 40, and 60 min 100 µl samples were pipetted onto filter discs which were washed twice, each time for ten min, in 10% and 5% trichloreacetic acid, and in ether-ethanol (1v:1v). After drying, radioactivity was counted in a toluene-based scintillator.

Preparation of Plasma Membranes and Binding of Labeled DIF

Cells were homogenized and centrifuged; the supernatant was separated on a discontinuous sucrose gradient, as described by *Forte et al.* [5]. Isolated membrane fractions were washed with phosphate-buffered saline (PBS) pH 7.4 and suspended there at a concentration of 50 µg in 50 µl. Labeled inhibitor, prepared as described below, was added in amounts ranging from 20 to 500 µl. The end volume in each test tube was adjusted to 550 µl. The mixtures were then incubated for 60 min at 30°C and centrifuged for 2 h at 100,000 g. The pellet was rinsed with ice-cold PBS and solubilized in Soluene.

Preparation of ^{3}H-Labeled Inhibitor

Two ml of the fraction with inhibitory activity after Sephadex and/or Dowex-separation were dialyzed with water and than lyophilized. The dry residue was dissolved in 200 µl of 0.1 M borate buffer, pH 8.5, and added immediately to 1 mCi of dry N-succinimidyl (2,3-^{3}H)propionate, with a specific activity of 68.8 Ci/mmol. After 1 h at 0°C and 13 h at 4°C, the mixture was diluted with 300 µl of PBS and the labeled inhibitor separated from unreacted material by chromatography on a Sephadex G-15 column, which was eluted with PBS.

Summary

The experiments presented here describe the separation of a factor from the maternal part of bovine placentas, capable of inhibiting protein and DNA synthesis in a specific manner. As a result, this factor inhibits the proliferation of tumor cells. The experiments further document a higher binding (and subsequent internalization) of this factor to receptors of the tumor-plasma membrane. This process presumably results in an increased inhibition in tumor cells of the synthesis of a protein, which is an essential component in the DNA-polymerase multienzyme complex assembly.

Zusammenfassung

In der vorliegenden Arbeit wird die Trennung eines Faktors aus dem maternen Anteil der Rinderplazenta beschrieben, der die Protein- und DNA-Synthese von Tumorzellen spezifisch hemmt. Die durchgeführten Bindungsstudien an Tumorzellmembranen ergaben eine erhöhte Bindung dieses Faktors an bestimmte Rezeptoren, die auf Tumorzellen gehäuft auftreten. Dieser Prozeß der Bildung eines Inhibitor-Rezeptor-Komplexes führt zu einer erhöhten Inhibition der Synthese eines Proteins, das essentiell für den DNA-Polymerase-Enzymkomplex ist.

References

1. Burke, J. F.; Duff, P. M.; Pearson, C. K.: Effect of drugs on desoxyribonucleic acid-synthesis in isolated mammalian cell nuclei. Biochem. J. *178:* 621 (1979).
2. Campisi, J.; Medrano, E. E.; Morreo, G.; Pardee, A. B.: Restriction point control of cell growth by a labile protein: Evidence for increased stability in transformed cells. Proc. Natl. Acad. Sci. USA *79:* 436 (1982).
3. DeLeo, A. B.; Lay, G.; Appella, E.; Dubois, G. C.; Lae, L. W.; Old, L. J.: Detection of a transformation-related antigen in chemically-induced sarcomas and other transformed cells of the mouse. Proc. natn. Acad. Sci. USA *76:* 2420 (1979).
4. Dippold, W. G.; Jay, G.; DeLeo, A. B.; Khoury, G.; Old, L. J.: p53 transformation-related protein: Detection by monoclonal antibody in mouse and human cells. Proc. natn. Acad. Sci. USA *78:* 1695 (1981).
5. Forte, J. G.; Forte, T. M.; Heinz, E.: Isolation of plasma membranes from Ehrlich-ascites-tumor cells. Biochim. Biphys. Acta *298:* 827 (1973).
6. Geipel, A.: Unterschiedliche biologische Wirkung des fetalen und maternen Anteils der Plazenta. Z. Gynäkol. *87:* 1433 (1965).
7. Gospodarowicz, D.; Moran, J. S.: Growth factors in mammalian cell culture. Ann. Rev. Biochem. *45:* 531 (1976).
8. Haas-Andela, H.: Die Wirkung von maternem Plazenta-Lyophilisat auf die Melanombildung bei lebend gebärenden Zahnkarpfen. Tagungsbericht, XXI. Jahrestagung über die zytoplasmatische Therapie, Stuttgart 1975.
9. Jacherts, D.; Jacherts, B.; May, G.: Prüfung der Wirksamkeit von Organextrakten an einem zellfreien System aus Hela-Zellen. Medsche Klin. *58:* 752 (1963).
10. Lau, C. C.; Pardee, A. B.: Mechanisms by which caffein potentiates lethality of nitrogen mustard. Proc. natn. Acad. Sci. USA *79:* 2942 (1982).
11. Letnansky, K.: Stoffwechselregulatoren der Plazenta und ihre Wirkung in Normal- und Tumorzellen. Exp. Path. *8:* 205 (1973).
12. Letnansky, K.: Faktoren aus der Plazenta, welche das Zellwachstum beeinflussen. I.: Hochmolekulare Faktoren. Österr. Z. Onkologie *2:* 31 (1974).
13. Letnansky, K.: Tumorspezifische Faktoren der Plazenta und Zellproliferation. Exp. Path. *9:* 354 (1974).
14. Letnansky, K.: Die Regulation der Zellproliferation in normalen und maligne entarteten Zellen; in Porcher, Theurer (eds), Biomimetik als Chance: Ein neues therapeutisches Prinzip, p. 103 (Enke, Stuttgart 1980).
15. Letnansky, K.: Versuche zur Charakterisierung eines tumorspezifischen Inhibitors aus dem mütterlichen Anteil der Rinderplazenta. Österr. Z. Onkologie *4:* 42 (1977).
16. Letnansky, K.: Nuclear proteins in genetically active and inactive parts of chromatin. FEBS letters *89:* 93 (1978).

17 Letnansky, K.: The phosphorylation of nuclear proteins in the regenerating and premalignant rat liver and its significance for cell proliferation. Cell Tiss. Kinet. 8: 423 (1975).
18 Letnansky, K.: Inhibition of thymidine incorporation into the DNA of normal and neoplastic cells by a factor bovine maternal placenta: Interaction of the inhibitor with cell membranes. Biosci. Rep. 2: 39 (1982).
19 Letnansky, K.: Early alterations in rat liver chromatin structure after a single dose of diethylnitrosamine. Biosci. Rep. 3: 185 (1983).
20 Medrano, E. E.; Pardee, A. B.: Prevalent deficiency in tumor cells of cycloheximide-induced cycle arrest. Proc. natn. Acad. Sci. USA 77: 4123 (1980).
21 Pardee, A. B.; Dubrow, R.; Hamlin, J. L.; Kletzien, R. F.: Animal cell cycle. Ann. Rev. Biochem. 47: 715 (1978).
22 Patt, L. M.; Houck, J. C.: The incredibly shrinking chalone. FEBS letters 120: 163 (1980).
23 Scatchard, G.: The attractions of proteins for small molecules and ions. Ann. N. Y. Acad. Sci 51: 660 (1949).
24 Theurer, K.; Triebel, R.: Verfahren zur Herstellung von Präparaten aus isolierten Plazentaranteilen. DBP 10 33 37 4.
25 Thompson, L. R.; McCarthy, B. J.: The effects of cytoplasmic extracts on DNA synthesis in vitro. Biochim. Biophys. Acta 331: 202 (1973).
26 Wrba, H.; Kalb, H. W.: Die spezifische Stoffwechselwirkung eines in der Plazenta enthaltenen Faktors in vitro. Naturwiss. 47: 85 (1960).
27 Wrba, H.: Krebsverhütung und Verhinderung der Krebsentstehung. Österr. Ärztezeitung 29: 1351 (1974).

Prof. Dr. rer. nat. K. Letnansky, Institut für Krebsforschung der Universität Wien, Borschkegasse 8a, A-1090 Wien (Austria)

Selective Effects of Sulfated Organ Lysates on the Clonal Growth of Normal Hematopoietic and Malignant Stem Cells in Vitro

H. R. Maurer

Institute for Pharmacy, Free University Berlin

Introduction

In addition to conventional cancer treatment by surgery, radio- and chemotherapy, an increasing number of drugs are administered that activate an immune response. These agents are called "biological response modifiers", since they alter the relationship between tumors and hosts by modulating the biological reaction of the host against its tumor with a resulting therapeutic effect. A great series of chemically defined and undefined substances of plant and animal sources belong to such agents; moreover, among them are several synthetic compounds. Sulfated organ lysates, prepared by *Theurer et al.*, for more than 25 years from various juvenile abattoir animals, contain macromolecular substances that showed significant tumor regressions in clinical and animal experiments [Review: Therapiewoche *33* (1983)]. The mechanism of these effects apparently involves a polyclonal stimulation of cytotoxic T-lymphocytes by the xenogenic components, particularilly polypeptides [*Munder*, 1983]. Moreover, direct inhibitions, in vitro, of the incorporation of ^{3}H-thymidine into melanoma-, Wish- and other tumor cells have been found [*Paffenholz and Theurer*, 1979; *Munder et al.*, 1982; *Letnansky*, 1983].

Since the latter method is problematic [*Maurer*, 1981] and highly sensitive tumor stem cell assays for screening of cytostatic drugs are now available [*Maurer*, 1983], dose-response relationships of the organ lysates (NeyTumorin®) were investigated using such assays. A significantly selective inhibition was observed of the tumor cells studied.

Materials and Methods

Materials

NeyTumorin® (manufacturer: VitOrgan Arzneimittel GmbH, 7302 Ostfildern 1, FRG), is a mixture of sulfated organ lysates from juvenile thymus, maternal placenta, liver, pancreas and other bovine and porcine organs. The mixture contains proteins, peptides, nucleic acids, lipids, polysaccharides, etc. The powder, in aliquots of 15 mg per 2 ml-ampoule, was dissolved in 0.9% NaCl before use and sterile-filtrated (0.22 µ). Batch used: No. 34264. Cis-platinum was purchased from Serva (Heidelberg, FRG), No. 19408.

Colony Assays

Granulocytic colonies (CFU-c) from mouse-bone marrow cells were cultured in agar-containing glass capillaries for 7 days, as described in detail [*Schupp et al., 1983*]. Colony-stimulating factor was prepared from mouse lung conditioned medium, as described [*Dietrich and Maurer, 1983*]. Peripheral, human T-lymphocytes were PHA-stimulated and grown to colonies for 7 days as described [*Meckert et al., 1983*]. Tumor cells were cloned for 7 days according to *Meckert and Maurer* (1983). For reviews on the colony assays see *Maurer* (1979, 1983). Medium for the granulocyte and lymphocyte colony assay: Dulbecco's modified Eagles' medium.

Tumor Cells

L 1210 (ATCC/CCI 219) from Flow, Ehrlich-ascites carcinoma (EAC, ATCC/CLL 77) from Prof. Dr. R. *Braun* (Marburg/Berlin, FRG), P 388 (ATCC/CCL 46) from Flow, M 1 from Dr. *Ichikawa* (Kyoto, Japan), L 929 (ATCC/CCL 1) from Flow, HL 60 (ATCC/CCL 240) from Dr. *Bombik* (Berlin). Medium for L 929: Dulbecco's modif. Eagle's Medium; for all other tumor cells: RPMI 1640.

A colony is defined as an aggregate > 40 cells of clonal origin.

Statistics

The mean and the standard error of the mean were calculated for each dose, based on the number of colonies per capillary (n = 3 capillaries).

Results

To assess the sensitivity of the colony assays for screening of cytostatic and cytotoxic substances, respectively, the effects of the cytostatic drug, cis-platinum, on the colony growth of normal myelopoietic stem cells and of various established tumor cells were studied. Figure 1 shows the dose-

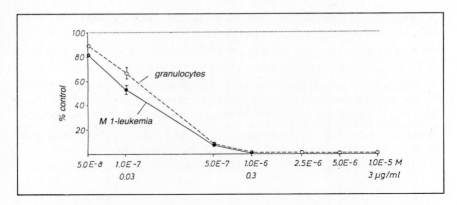

Fig. 1. Effect of cis-platinum on the colony growth of mouse granulocytes (100% control = 20 ± 1 colonies) and M-1 leukemia cells (100% control = 47 ± 1 colonies). 1.0E-7 = 1 × 10^{-7}M (E = exponent).

response for normal mouse granulopoietic colonies following stimulation by the colony-stimulating factor (CSF) and for a myeloid leukemia (M 1) without any additional stimulator: 10^{-7} M cis-platinum (= 0.03 µg/ml) inhibit colony growth of granulocytes by about 30%, of the leukemic cells by about 45% relative to untreated controls. The M-1 leukemia is a sensitive cell line that can be activated for differentiation by endogenous and synthetic compounds of various kinds, and has proved suitable for the

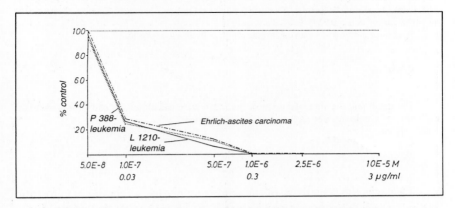

Fig. 2. Effect of cis-platinum on the colony growth of Ehrlich-ascites carcinoma cells (100% control = 40 ± 1 colonies), P 388- (100% control = 41 ± 0.4 colonies) and L-1210 leukemia cells (100% = 55 ± 1 colonies).

investigation of in-vitro phenomena of differentiation [*Ichikawa*, 1969; *Hozumi*, 1983]. The lympahtic mouse-leukemias, L 1210 and P 388, have been and are still extensively used world-wide, to screen various natural and synthetic products for cytotoxicity. Both cell lines, as well as the long-established Ehrlich-ascites carcinoma, reveal a comparably high sensitivity for cis-platinum (fig. 2): 10^{-7} M inhibit growth by about 75%. These relatively high sensitivities of normal and malignant cells for cis-platinum, demonstrate the validity of the colony assays for in vitro screening purposes.

In contrast to cis-platinum, NeyTumorin® shows significant inhibitions, mostly at concentrations of > 100 µg/ml, suggesting a relatively weak cytotoxicity, if any. However, considerable differences are noteworthy in the dose-responses of the different colony assays: colonies of PHA-stimulated, human, peripheral T-lymphocytes are inhibited starting only at about 800 µg/ml by 40% and of CSF-stimulated granulocytic colonies starting only at about 700 µg/ml by 40% (fig. 3). The corresponding ID_{50}-values for both cell lines are about 900 µg/ml. Thus, the growth of these normal cells is only inhibited by relatively high concentrations of NeyTumorin®, which may be due to an unphysiological increase of the osmotic pressure and an oversaturation of the medium by peptides, polysaccharides, etc. However, the L-1210 leukemia already responds at 200 µg/ml, the P-388 leukemia at 100 µg/ml NeyTumorin®; corresponding ID_{50}-values: 400 µg/ml and 270 µg/ml, respectively. It seems unlikely, that the significant differences result from different growth parameters of normal and

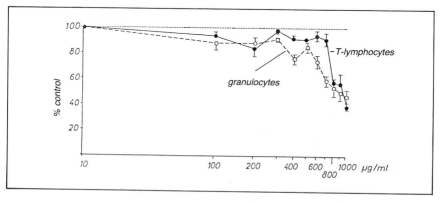

Fig. 3. Effect of NeyTumorin® on the colony growth of human T-lymphocytes (100% control = 42 ± 0.3 colonies) and mouse granulocytes (100% control = 22 ± 1 colonies).

malignant cells, since all values relate to the corresponding controls and all colony numbers range in a similar order of magnitude.

Moreover, the myeloid M-1 leukemia and the Ehrlich-ascites carcinoma, widely used for cytostatic drug-screening, respond to NeyTumorin® beginning at about 200 µg/ml (fig. 5); corresponding ID_{50}-values: about 500 µg/ml and 470 µg/ml, respectively.

The human promyelocytic HL-60 leukemia is quite sensitive to NeyTumorin®: 10 µg/ml causes about 15%, 100 µg/ml about 50% inhibition of colony growth (fig. 6). The HL-60 leukemia is preferably used to study

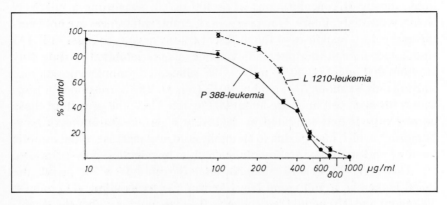

Fig. 4. Effect of NeyTumorin® on the colony growth of L-1210 (100% control = 41 ± 0.3 colonies) and P-388 leukemia cells (100% control = 46 ± 1 colonies).

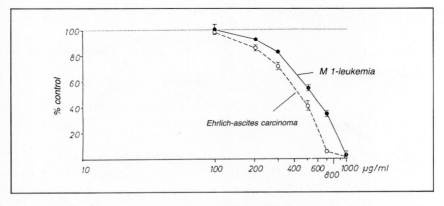

Fig. 5. Effect of NeyTumorin® on the colony growth of M-1 (100% control = 42 ± 1 colonies) and Ehrlich-ascites carcinoma cells (100% control = 42 ± 1 colonies).

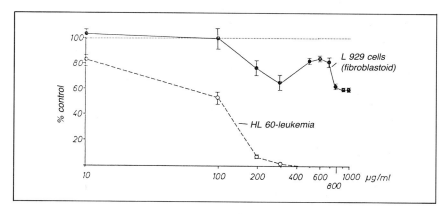

Fig. 6. Effect of NeyTumorin® on the colony growth of fibroblastoid L-929 cells (100% control = 50 ± 1 colonies) and HL-60 promyelocyte leukemia cells (100% control = 24 ± 0.4 colonies).

differentiation phenomena, as it can be stimulated to differentiate by various agents within the myelopoietic series [*Collins*, 1978; *Hozumi*, 1983].

In comparison, the fibroblastoid L-929 cell line shows a relatively moderate inhibition (fig. 6): A significant inhibition is seen in 300 µg/ml ($ID_{50} \triangleq 35\%$), which is reduced at 500–600 µg/ml ($ID_{50} \triangleq 10\%$), but increased at 800 µg/ml ($ID_{50} = 40\%$). This odd dose-response relationship is reproducible. It should be noted, that the L-929 cell line, from mouse connective tissue origin, is often used for carcinogenesis and toxicity studies. It produces sarcomas only in irradiated C3H/HeN mice, i.e., it reveals only a relatively "weak" malignancy.

Discussion

The results reveal considerably different effects of NeyTumorin® on
(a) normal, i.e. non-malignant cells (mouse granulocyte, human T-lymphocytes, fig. 3),
(b) cells of "moderate" malignancy (L-929 cells, fig. 6) and
(c) cells of "advanced" malignancy (M-1, P-388, L-1210, HL-60 leukemia and Ehrlich-ascites carcinoma (fig. 4 and 5).
It is remarkable that the growth of all tumor cells studied is significantly more inhibited than that of non-malignant cells (factor 1.6 to about

9), which cannot be inhibited up to 100% even by a dose as high as 1 mg/ml. This also holds true for the "moderately" malignant L-929 cells (fig. 6). These quantitative results confirm earlier findings by *Paffenholz and Theurer* (1979) and *Munder et al.* (1982), that ^{3}H-thymidine uptake by melanoma- and Wish cells can be inhibited up to 50% by the addition of 10% NeyTumorin® to the culture medium. To assess cell proliferation, the determination of colony growth shows an essential advantage, compared with the ^{3}H-thymidine method, in that it omits the various potential artifacts that may render results obtained with this method meaningless [*Maurer*, 1981]. Moreover, colony assays are increasingly favored for screening purposes [*Salmon and von Hoff*, 1981; *Yarbro*, 1981], since they are based on the widely accepted theory of the clonal evolution of tumors. The high sensitivities of the stem cell assays are validated by the cis-platinum results (fig. 1) which largely surpass those of corresponding ^{3}H-thymidine tests (unpublished results) confirming previous observations on a comparison of sensitivities of colony- and ^{3}H-thymidine assays [*Osman and Maurer*, 1980].

Besides the known indirect inhibitory effect of NeyTumorin®, which may be explained by a polyclonal activation of cytotoxic T-lymphocytes by xenogenic components of NeyTumorin® [*Munder*, 1983], the mixture of sulfated organ lysates from placenta, juvenile thymus, liver, and other organs, exerts an apparently selective inhibition on the tumor cells studied. Its mode of action is still unknown. However, on the basis of the dose-response relationships, the inhibitory effects cannot be explained by mere cytotoxicity.

The following questions may be asked to elucidate the inhibitory effect:

(1) How far can the apparent selective inhibition of tumor cell growth be extrapolated to other malignant cells? Does the cytotoxicity correlate, in general, with the degree of malignancy of the tumors, which is suggested by the experiments described, but not evidenced beyond any doubt?

(2) Does the inhibition involve direct, selective killing of growing tumor cells or retardation of colony growth?

(3) How do the kinetics of inhibition develop?

(4) Is the inhibition reversible and, if so, for how long?

(5) Which phase of the cell cycle is affected by NeyTumorin®? Is there a difference from most cycle-specific, conventional cytostatics?

(6) Do the inhibitors contained in NeyTumorin® attack the tumor cell

membrane, as suggested by the observations of *Letnansky* (1982) with bovine placenta extracts?

(7) Are there common or different properties among the agents of NeyTumorin® and other inhibitors, e.g., chalones from placenta [*Letnansky,* 1974], thymus and liver?

(8) To which biochemical class do the inhibitors of NeyTumorin® belong?

(9) Does NeyTumorin® contain substances that stimulate the differentiation of tumor cells (e.g., M-1 and HL-60 cells)?

(10) Does NeyTumorin® contain polyamines in free or bound form? Polyamines may produce cytotoxic polyaminoaldehydes in the presence of polyaminooxidases [*Allen,* 1983].

It may be assumed on the basis of the findings of *Letnansky* (1983), *Munder* (1983), *Paffenholz and Theurer* (1979), that the inhibitors are contained in NeyTumorin® in minute quantities. Following biochemical isolation, substances may be available which, in concentrations similar to those of peptide hormones (10^{-9}–10^{-11} M), may reveal no toxicity and no severe, undesired side-effects known as major drawbacks of conventional cytostatic drugs (particularly on lympho- and granulomonopoiesis). This most favorable advantage should justify an early and intensive investigation to isolate and purify the inhibitor(s) in order to solve the questions mentioned above.

Acknowledgment

The most skilful technical assistance of Mrs. C. Meckert is gratefully acknowledged.

Summary

The effects of sulfated organ lysates from juvenile thymus, maternal placenta, liver, and other bovine and porcine organs (NeyTumorin®) were investigated in vitro using clonal stem cell assays with mouse granulocytes, human T-lymphocytes and established tumor cell lines (L-1210, P-388, M-1, H-60 leukemia, Ehrlich-ascites carcinoma and fibroblastoid L-929 cells). Dose-response curves indicated that the growth of all malignant cell-lines investigated is inhibited significantly (factor 1.6–9) more than that of normal granulocytes and T-lymphocytes. L-929 cells were much less inhibited. A selective growth inhibition can be deduced from the tumors studied. The validity and high sensitivity of the stem cell assays for screening purposes is demonstrated by comparison with cis-platinum effects.

Zusammenfassung

Die Wirkungen eines sulfatierten Organlysats aus juvenilem Thymus materner Plazenta, Leber u. a. Organen vom Rind und Schwein (NeyTumorin®) wurden in klonalen Stammzell-Assays mit Mäuse-Granulozyten, menschlichen T-Lymphozyten und etablierten Tumorzellen (L-1210-, P-388-, M1-, HL-60-Leukämie, Ehrlich-Ascites-Karzinom und fibroblastoiden L-929-Zellen) in vitro untersucht. Die Dosis-Wirkungsbeziehungen lassen erkennen, daß das Wachstum aller untersuchten Tumorzellen durch das Präparat signifikant stärker gehemmt wird (Faktor 1,6–9) als das der normalen Granulozyten und T-Lymphozyten. L-929-Zellen werden bedeutend schwächer gehemmt. Damit ergibt sich für die untersuchten Tumoren eine deutlich selektive Wachstumshemmung. Die Brauchbarkeit und hohe Empfindlichkeit der Stammzell-Assays für Screening-Zwecke wurde im Vergleich mit Cis-Platin aufgezeigt.

References

Ali-Osman, F.; Maurer, H. R.: Comparison of cytostatic sensitivities of L 1210 cells and human stimulated lymphocytes in three cell proliferation assays. J. Cancer Res. Clin. 98: 221–231 (1980).
Allen, J. C.: Biochemistry of the polyamines. Cell Biochem. & Function 1: 131–140 (1983).
Collins, S. J.; Ruscetti, F. W.; Gallagher, R. E.; Gallo, R. C.: Terminal differentiation of human promyelocytic leukemia cells induced by dimethylsulfoxide and other polar compounds. Proc. natn. Acad. Sci., USA 75: 2458–2462 (1978).
Dietrich, C.; Maurer, H. R.: Herstellung von Colony-Stimulating Factor (CSF) aus Mäuselungen-conditioniertem Medium (MLCM); in Maurer (ed.), Manual "Zellkultur-Methoden", pp. 103–111 (Selbstverlag, Berlin 1983).
Hozumi, M.: Fundamentals of Chemotherapy of Myeloid Leukemia by Induction of Leukemia Cell Differentiation. Adv. Cancer Res. 38: 121–169 (1983).
Ichikawa, Y.: Differentiation of a cell-line of myeloid leukemia. J. Cell Physiol. 74: 223–234 (1969).
Letnansky, K.: Tumorspezifische Faktoren der Placenta und Zellproliferation. Exp. Path. 9: 354–360 (1974).
Letnansky, K.: Inhibition of thymidire incorporation into DNA of normal and neoplastic cells by a factor from bovine maternal placenta: Interaction of the inhibitor with cell membranes. Bioscience Rep. 2: 39–45 (1982).
Letnansky, K.: Entdeckung zellulärer Rezeptoren für antitumorale plazentare Faktoren in NeyTumorin. Therapiewoche 33: 59–61 (1983).
Maurer, H. R.: In vitro colony growth of granulocytes, macrophages, T- and B-lymphocytes in agar capillaries. Acta haemat. 62: 322–325 (1979).
Maurer, H. R.: Potential pitfalls of ^{3}H-thymidine techniques to measure cell proliferation. Cell & Tissue Kinet. 14: 111–120 (1981).
Maurer, H. R.: Prüfung potentieller tumorhemmender Wirkstoffe mit dem Tumorstammzell-Assay in vitro: Therapiewoche 33: 44–50 (1983).
Meckert, C.; Maurer, H. R.: Klonierung von Tumorzellen in Agar-haltigen Glaskapillaren (Mikro-Tumorstammzell-Assay); in Maurer (ed.), Manual "Zellkultur-Methoden", pp. 117–123 (Selbstverlag, Berlin 1983).
Meckert, C.; Echarti, C.; Maurer, H. R.: Kultivierung von T-Lymphozyten-Kolonien aus menschlichem Vollblut in Agar-haltigen Glaskapillaren; in Maurer (ed.), Manual "Zellkultur-Methoden", pp. 112–116 (Selbstverlag, Berlin, 1983).

Munder, P. G.; Stiefel, Th.; Widmann, K. H.; Theurer, K.: Antitumorale Wirkung xenogener Substanzen in vivo und in vitro. Onkologie 5: 98–104 (1982).

Munder, P. G.: Experimentelle Untersuchungen über den antitumoralen Wirkungsmechanismus von NeyTumorin. Therapiewoche 33: 71–73 (1983).

Paffenholz, V.; Theurer, K.: Einfluß von makromolekularen Organsubstanzen auf menschliche Zellen in vitro: II. Tumorzellkulturen. Kassenarzt 19: 1976–1887 (1979).

Salmon, S. E.; von Hoff, D. D.: In vitro evaluation of anticancer drugs with the human tumor stem cell assay. Semin. Oncol. 8: 377–385 (1981).

Schupp, D.; Königstein, M.; Maurer, H. R.: Kultivierung von Granulozyten-Kolonien aus Mäuse-Knochenmark-Stammzellen in Agar-haltigen Glaskapillaren; in Maurer (ed.), Manual "Zellkultur-Methoden", pp. 103–111 (Selbstverlag, Berlin 1983).

Yarbro, J. W. (ed.): Clinicals Trials: Design and Analysis. Semin. Oncol. 8: 4 (1981).

Prof. Dr. rer. nat. H. R. Maurer, Institut für Pharmazie der Freien Universität Berlin, Königin-Luise-Straße 2 + 4, 1000 Berlin 33 (Germany)

Studies of the Stimulation of Non-Specific Defense Mechanisms by NeyTumorin®-Sol

A. Mayr, M. Büttner, S. Pawlas

Institute for Medical Microbiology and Infectious and Epidemic Diseases, Faculty of Veterinary Medicine of Ludwig-Maximilian University, Munich, FRG

Introduction

The starting and growth of tumors is influenced to a major extent by the performance of the body's own resistance. An important role is played by non-specific defense mechanisms in the case of both immunogenic and non-immunogenic tumors [11].

Via receptors, it is possible for certain effector cells to be bound to transformed cells, i. e., cells which in their antigen structure are not or no longer accepted by the organism, and for then to lyse the latter. Natural (spontaneous) killer cells and cytotoxic macrophages form the basis of this effector cell population. In the course of an immunogenic tumor process, the organism affected is able to resist tumor cells in an immunological manner. Antibodies (IgG) are formed against certain surface-structure proteins of tumor cells enabling another type of effector cell, the killer or null cell, to bind to the tumor cell, thereby causing its destruction.

Certain endogenous mediators, such as interferon or interleukin, stimulate these defense manifestations or in some cases, as indispensable transmitter substances, even start them going.

However, depending on the individual functional state of the defense system, it can happen that mediators predominate, which stimulate the suppressor cells and which may be produced or released by the tumor itself [3]. It is known that non-specific defense mechanisms can be medicinally enhanced in respect to tumor cells by the NeyTumorin®-Sol preparation [2, 7, 9, 10].

The mechanism of action of this ‚stimulator' is presently being intensively studied in vitro and in vivo and here, to a certain extent, already in therapy.

Some aspects of non-specific defense intensification by NeyTumorin®-Sol have been studied in our institute.

The first requirement, that such medicaments must meet before their efficacy can be tested with various biological tests, is that they do not cause damage. Toxicological and pharmacological investigations showed that NeyTumorin®-Sol is pyrogen-free and non-mutagenic [4].

Material and Methods

In Vivo Tests

To demonstrate its harmlessness, NeyTumorin®-Sol was administered to baby mice, 1–2 days old, as an undiluted active agent in a quantity of 0.1 ml by the subcutaneous and intraperitoneal routes. The animals were clinically examined daily over a period of 14 days for possible intolerance reactions.

A test for a non-specific defense intensification was carried out in the infection-stress model with the "VSV-test"; baby mice 1–2 days old, were subcutaneously or intraperitoneally pretreated with NeyTumorin®-Sol (0.1 ml active agent undiluted and diluted 1:4) and exposed 24 h later by the intraperitoneal route to the rhabdovirus of vesicular stomatitis (10–50 LD_{50}), which is highly virulent for the animals.

An induction of serum interferon was checked by two-times administration of 0.5 ml active substance to adult NMRI-mice by the intraperitoneal route. The indirect demonstration of interferon has been carried out by the plaque reduction test (PRT). The vesicular stomatitis virus strain Indiana, set to 30 plaque-forming units, served as the test virus and a permanent mouse-line LTK (Subcutis) was used as the indicator cell culture. The non-inactivated serum of 5 mice was pooled 8, 24, 48, and 96 h after the application of NeyTumorin®-Sol and then tested in dilutions (strength) for its interferon content in the PRT (in vitro).

In Vitro Tests

In the direct plaque-reduction test, the preparation per se was tested for an in vitro interferon induction in the cell culture system. For this, dilutions to the strength 2 were prepared and 0.1 ml each placed on cell culture plates (dense cell lawns). After a 4-h incubation at 37°C, the substance dilutions (in the indirect test the serum dilutions) were removed by suction and each cell culture inoculated with 0.1 ml test virus in the working dilution. After 1-h incubation and removal of the test virus the cultures were overlayed with agar and incubated for further 48 h. Then they were stained with neutralred and kept at room temperature overnight. The test was evaluated by counting the plaques (mean values from 4 identical cultures) and comparison with the virus control.

The influence of NeyTumorin®-Sol on resistance through the intermediary of the cell has been checked by the lymphocyte transformation test by an increase in the spontaneously cell-communicated cytotoxicity and/or the phagocytosis capability of leukocytes in vitro. The effector cells used for this were taken exclusively from the peripheral blood of pigs.

The lymphocyte transformation test was carried out by the usual method using porcine lymphocytes, labeled with ^{3}H-thymidine and separated by Ficoll plaque centrifuging. PHA (10 µg/ml) was used as a control stimulator. NeyTumorin®-Sol was added undiluted 1:2 and 1:4 to the total lymphocyte fraction. The substance was tested alone and in combination with PHA. The evaluation was carried out via the ^{3}H-thymidine incorporation rate in counts per min in the liquid scintillation counter [8].

The spontaneous cytotoxicity by cell mediation of peripheral blood cells was tested with porcine lymphocytes, separated by Ficoll plaque centrifuging, in micro-plates. 10 µl Ney-Tumorin®-Sol, diluted in the strength of 1, were added to each of the effector-target cell-incubation mixtures 100:1/50:1/25:1 and 12:1 (identical triplets) directly with the addition of the Cr^{51}-labeled Vero target cells (monkey kidney line). A chromium-51 release test was then carried out for 16 h at 37°C, at the end of which the effector cell activity was measured via the gamma radiation of the Cr^{51} released from lysed target cells calculated in percent of the specific lysis [1].

The phagocytosis capability of porcine leukocytes (7×10^6/ml) was assessed after the addition of a suspension of Baker's yeast cells (7×10^6/ml NaCl sol). The same volumes (62 µl each) of leukocyte suspensions, yeast-cell suspensions, Hank's buffer solution and the serum of the same animal as a complementary source, were incubated for 1 h at 37°C, vigorous shaking being maintained. During the incubation time, NeyTumorin®-Sol acted on the leukocytes undiluted and diluted in powers of 10 up to 10^{-6}. After the lysing of the leukocytes with sodium deoxycholate (2.5% in aqua dest., pH 8.7), the dead, phagocytized yeast cells were stained by the addition of 1 ml 0.01% methylene blue solution. The proportion of dead yeast-cells in the total cell count was determined in a haemocytometer. Preparations without leukocytes served as controls; 300 yeast cells were counted in each case [5].

Results

In Vivo Tests

No clinical signs of intolerance at all were found in any of the baby mice after the application of NeyTumorin®-Sol, neither i.p. (30 animals) nor s.c. (27 animals). Three of 30 animals (undiluted substance) and 3 of 26 animals (1:4-dilution) survived the infection-stress after an intraperitoneal pretreatment with NeyTumorin®-Sol; the mortality in the placebo group being 100%. No protection was induced by the subcutaneous treatment. To a slight extent, we were able to demonstrate an induction of serum interferon in adult mice with the plaque reduction test (12% inhibition with 1:2-serum dilution). Only in the sera obtained in the 24 h after the i.p. application of NeyTumorin®-Sol was it possible to demonstrate interferon.

In Vitro Tests

In the direct plaque-reduction test, the preparation had no effect on the cell system in the sense of a stimulation with an influence on the local virus proliferation (plaque reduction).

In the lymphocyte transformation test, NeyTumorin®-Sol did not lead to any significant stimulation indices. When added to the lymphocytes simultaneously with phytohaemagglutinin (10 µg/ml), the stimulating effect of PHA was completely depressed.

As proof of a non-specific increase in the cytotoxicity (SCC), NeyTumorin®-Sol achieved a specific lysis of 10% in the 1:1000-dilution and of 11% in the 1:10 000-dilution in the Cr^{51} release test. In comparison with the specific lysis of untreated lymphocytes from the same animal with the same effector/target-cell ratio (100:1) of 9%, these values must be assessed as a scarcely detectable increase. However, in the dilutions of 10^{-2}, 5×10^{-2} and 10^{-3}, the preparation was still able to stimulate specific lysis values of 1 to 3% with a very low effector-cell count (ratio 12:1) in identical test cultures, whereas untreated lymphocytes from the same animal were inactive.

The phagocytosis capability of porcine leukocytes remained uninfluenced after exposure to NeyTumorin®-Sol.

Discussion of the Results

The medicamentous stimulation of the body's own non-specific resistance is increasingly gaining importance in the prophylaxis and therapy of tumors. In the very recent English-language literature on the subject, substances suitable for this have been designated as "biological response modifiers". We also have drawn attention to the stimulation of non-specific resistance since 1979 and coined the term of "Paramunisierung" ('paramunization') for this, which was intended to indicate that in addition to ('para') specific immune mechanisms, there are also non-specific defense mechanisms which are active in an organism, giving a non-specific protection which helps the organism to ward-off exogenous noxae before the development of an immunity or in parallel with this [6].

In the demonstration of 'paramunizing' activities of 'biological response modifiers' or 'paramunity inducers', the results in clinical practice and in controlled laboratory experiments are often diametrically opposed to each other. The reason for this is that in the clinical sphere even slight rises in the defense mechanisms have a positive effect on the course of diseases. However, they are scarcely detectable in laboratory experimentation under difficult conditions.

NeyTumorin®-Sol has proved its value in clinical practice in a variety of ways. It was, therefore, of interest to ascertain whether NeyTumorin® also has a stimulating activity on the non-specific body defense in strictly controlled laboratory tests. The laboratory models used by us for this purpose are extremely exacting in their requirements. In this connection, we have investigated in our models numerous inducers of 'proven' value in clinical practice. The results were always very unsatisfactory. It was, therefore, all the more surprising that with NeyTumorin® we were able to demonstrate stimulating activities; both in the interferon and in the NK model, slight though these were. Both activities are important for tumor resistance. The greatest importance in this is attributed to the stimulation of spontaneous cytotoxicity mediated by effector cells. This is the first effective defense mechanism for the limitation of immunogenic and non-immunogenic tumors. It is possible that the good results with NeyTumorin®-Sol in clinical practice are to be partly attributed to this mechanism of action.

Summary

NeyTumorin®-Sol was tested in various laboratory models with respect to a stimulant effect on various non-specific defense mechanisms. An activation of the spontaneous cytotoxicity by cell mediation and a transitory interferon induction were demonstrated. For the clinical efficacy of NeyTumorin®, the stimulation of the spontaneous cytotoxicity by cell mediation is possibly of particular significance.

Zusammenfassung

NeyTumorin®-Sol ist in verschiedenen Labormodellen bezüglich stimulierender Wirkung auf verschiedene unspezifische Abwehrmechanismen überprüft worden. Nachgewiesen wurde eine Aktivierung der spontanen, zellvermittelten Zytotoxizität und eine kurzfristige Interferoninduktion. Für die klinische Wirksamkeit von NeyTumorin® besitzt möglicherweise die Stimulierung der spontanen, zellvermittelten Zytotoxizität eine besondere Bedeutung.

References

1 Büttner, M.: Stimulierung der natürlichen (spontanen) zellvermittelten Zytotoxizität mit verschiedenen Inducern. Fortschr. Vet. Med. *37:* 36–41 (1983).
2 Dörr, H. W.: Erfahrungen mit der Organo- und Immuntherapie bei inoperablen Tumoren und in der Nachsorge. Erfahrungsheilkunde *3:* 156–163 (1983).
3 Engler, H.; Kirchner, H.: Natürliche Killerzellen. Med. Mo. Pharm. *1:* 3–10 (1983).
4 Gillissen, G., 1983: Toxikologisch-pharmakologische Untersuchungen mit NeyTumorin®-Sol. Conference: Cytoplasmatic Therapy, Stuttgart, 1983.
5 Lehrer, R. J.; Cline, M. J.: Interaction of Candida albicans with human leucocytes and serum. J. Bacteriol. *98:* 996–1004 (1969).

6   Mayr, A.; Raettig, R.; Stickl, H.; Alexander, M.: Paramunität, Paramunisierung, Paramunitätsinducer. Fortschr. Med. *97:* 1159–1165 (1979).
7   Munder, P. G.: Experimentelle Untersuchungen über den antitumoralen Wirkungsmechanismus von NeyTumorin®-Sol. Therapiewoche *33:* 71–73 (1983).
8   Pawlas, S.: Untersuchungen zur Charakterisierung der Lymphozytenpopulation von Schwein und Rind. Vet. med. thesis, Munich, 1979.
9   Stiefel, T.: Intravenös-gängiges NeyTumorin®-Sol. Therapiewoche *33:* 74–78 (1983).
10 Theurer, K. E.: Die zytoplasmatische Tumortherapie. Krebsgeschehen *1:* 12–15 (1983).
11 Warner, J. F., and G. Dennert: Effects of a cloned-cell line with NK activity on bone marrow transplants, tumor development and metastasis in vivo. Nature *300:* 31–34 (1982).

Prof. Dr. Dr. h. c. mult. A. Mayr, Institut für Medizinische Mikrobiologie, Infektions- und Seuchenmedizin, Veterinärstr. 13, D-8000 München 22 (FRG)

Pilot Study on the Influence of a Biological Response Modifier (NeyTumorin®) on the Plasma Membrane of Human Tumor Cells (Wish) in Vitro – in Comparison with a Chemocytostatic Agent (6-Mercaptopurine)

U.-P. Ketelsen

University Children's Clinic and Electron Microscopy Department of the Max-Planck Institute for Immunobiology, Freiburg i. Brsg., FRG

Comprehensive conventional ultrastructural studies of the ultra-thin section of tumor cells in vitro have, so far, proved unsatisfactory in respect to the tumor-specificity of individual cell structure changes, i.e., no constant morphological abnormality, not even a single cytoplasmic cell structure, could be associated with the malignancy of the tumor cell. However, conventional and scanning electron-microscopy examinations of tumor cells from the cell culture, frequently show an unusual cell-surface activity, which is expressed in the increased formation of cell processes, microvilli or tiny, blister-like distensions of the cell membrane.

The Cell Membrane – Counterpart of the Genome

The cell membrane not only has already well-known structural and metabolic functions, but also acts, as we now know, as a regulative center which transforms signals from the micro-environment of the cell and thus triggers certain consequent reactions in the cell. The outer cell membrane can now be defined, in the broadest sense, as a genuine counterpart of the genome, in that it exercises the functions of the exchange of information with the environment. While the genome, on the other hand, is responsible for storage of information.

In the model concept now generally accepted, phospholipids are arranged in the outer cell membrane as a double planar layer, in which peripheral or integral structural and transport proteins are embedded mosaic-like and asymmetrically. They are able to move in the lipid matrix. The model of the "fluid mosaic membrane", first postulated by *Singer and Nicolson* [6] is very simplified since cells display a much greater stability than could be achieved by a double-lipid layer alone, where proteins are freely and completely mobile. This stability is partly achieved by peripheral proteins, which are found only in the hydrophilic area of the membrane and are associated with several integral membrane proteins, and with these, form a supporting structure. An example of this is the spectrin on the inner side of the erythrocyte membrane. Furthermore, filament structures are associated with the membrane in the form of the cytoskeleton, which also appears to control the mobility of some of the integral membrane proteins. The membrane model developed from morphological, biochemical, and biophysical results is confirmed by ultrastructural studies with the aid of the freeze-fracture method (for method see [3]). With the freeze-fracture, the objects fracture along preferred weakness zones. In the case of membranes, this is the hydrophobic middle zone [2]. The freeze-fracture thus displays inner views of membranes. Outer-split membrane halves, designated by "EF", and inner-split membrane halves, designated by "PF" are thus obtained from each specimen. On these split halves, the integral membrane proteins appear in the smooth lipid matrix as particles, more numerous on the inner-split halves than on the outer ones. This characterizes the membrane's asymmetry in the morphological aspect, as well.

Membrane Changes in Invasive Carcinomas

The invasive properties of tumor cells are determined by numerous factors, of which the cell-surface modification appears to play a very important role. In addition to the information and metabolic functions briefly outlined here, the cell membrane in tissues such as the epithelium or endothelium, makes possible the coherence of the individual cells via specialized contact points (gap junctions, desmosomes). In recent years, it has been demonstrated in vivo that changes in these contact points precede the development of invasive carcinomas [4]. It is obvious, however, that the membrane changes in the preneoplastic and neoplastic states are not only related to these contact points alone, but are also correlated with changes in the whole of the molecular architecture of the plasmalemma [8].

In the following pilot study, we have examined normal human amnion cell cultures with human Wish amnion tumor cells after 19 culture days in the ultra-thin section and with the aid of the freeze-fracture method. In addition, the same methods were used to analyze Wish cell cultures which were treated with 6-mercaptopurine and, in parallel, with NeyTumorin®.

The freeze-etching of the cell cultures treated with 6-mercaptopurine was carried out after a 6-day treatment-period as sufficient cell material for this special study was no longer available at this time because of progressive cell degeneration. Cultures treated for 10 days with 6-mercaptopurine were merely studied in ultra-thin sections. The cell cultures treated with NeyTumorin were analyzed after a treatment-period of 10 and 17 days by the freeze-fracture technique, in ultra-thin sections.

The examination of the cell membranes in the freeze-fracture was concentrated on

(1) the number of intramembranous particles;

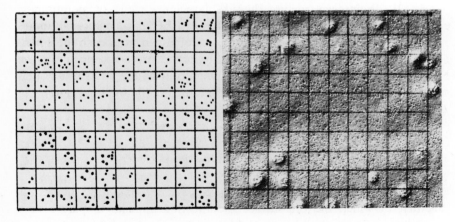

Fig. 1. Quantitative analysis of a cell membrane fracture-face as obtained with a skeletal muscle cell membrane (see *Ketelsen,* 1980). Test-grid (transparent foil) with 100 squares over the electron microscopic picture of a fractured surface of the plasmalemma (P-face). Next to it, example (not identical with the adjacent picture) of intramembranous particles marked as points on the test-grid (recording the number of points by morphometer and calculation of their topical distribution by on-line computer).

(2) the two-dimensional topography or distribution within the membrane.

A laboratory assistant, who had no code knowledge relating to the various cell cultures, carried out the evaluation of 7 to 15 P- and E-faces of the split-membrane halves for each cell preparation. The evaluation was made with electron-microscopic pictures with a final magnification of × 60 000.

The number of intramembrane particles and their topographical distribution was determined in membrane areas of 1 µm^2 with the aid of a test-grid (fig. 1) with a 6 cm edge-length, demarcating the size of the test-area. This test-grid is divided into 100 squares of 36 mm^2 each. The quantitative analysis was carried out with a semi-automatic Leitz morphometer and evaluated with a programmable on-line computer. All the particles within the test-area were pencil-marked on the plotting panel of the Leitz instrument. The computer now recorded the number and the topographical distribution of the particles within the whole of the test-area. The topographical distribution was expressed in the form of the coefficient of dispersion [5]. Values above 1.4 indicate a statistically significant aggregation, whereas values below 0.67 point to a highly-ordered organization and values between 0.67 and 1.4 indicate a random distribution of the intramembrane particles [5].

In ultra-thin sections, the normal amnion cell in vitro has an epitheloid character with a few microvilli, a well-formed, rough endoplasmic reticulum, mitochondria with a regular structure, nuclei with finely dispersed chromatin and nucleoli, some of which are prominent (fig. 2a). Near the plasma membrane, delicate filament bundles are often demonstrated which correspond to the tonofilaments of epithelial cells in vivo (fig. 2b). We also observed cell-to-cell contacts, analogous to so-called gap junctions. In the freeze-fracture, the split membrane halves have diffusely dispersed particles in the lipid base matrix (figs. 2c and d); these numbering 693 per µm^2 as a mean value on the P-faces and 284 particles per µm^2 on the E-faces. The coefficient of dispersion displays a random distribution of particles without a significant tendency to aggregation (table I).

While in the ultra-thin section the untreated Wish cell differs little in its ultrastructure from the normal amnion cell, apart from an accumulation of microvilli (figs. 3a and b), clear differences can be demonstrated with the aid of the freeze-fracture method in the molecular architecture of the plasmalemma. In the freeze-fracture of the plasmalemma of untreated

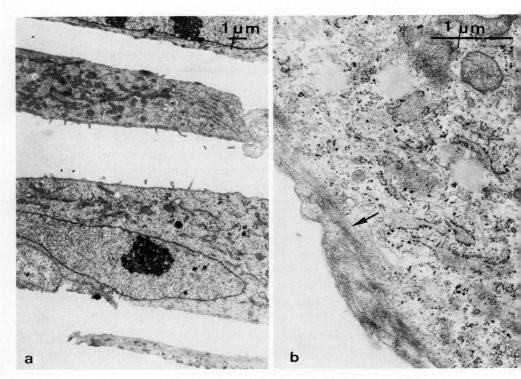

Fig. 2, a–d. Normal human amnion cells in vitro (19 culture-days). (a) and (b) Amnion cells in ultra-thin sections with only sporadic, small folds of the cell membrane, well-developed, rough endoplasmic reticulum and nuclei with finely dispersed chromatin and, to some extent, prominent nucleoli (a). With greater magnification (b), delicate filament bund-

Wish cells, there is a significant increase in the integral membrane particles, especially on the P-faces (fig. 3c) with a mean value of 1414 particles per μm^2. In comparison with the normal check, the coefficient of dispersion indicates an aggregation tendency for the particles in the membrane (table I). In ultra-thin section and after a 6-day treatment with 6-mercaptopurine, Wish cells display a marked formation of microvilli, a close-packed cytoplasm, large nuclei with prominent nucleoli and an enlarged, rough endoplasmic reticulum (fig. 4a). Wish cells which were treated for 10 days with 6-mercaptopurine (1.5×10^{-5} g) display marked degenerative changes. Their nuclei are already swollen in a blister-like manner which is evident under the light microscope. The chondriome is charac-

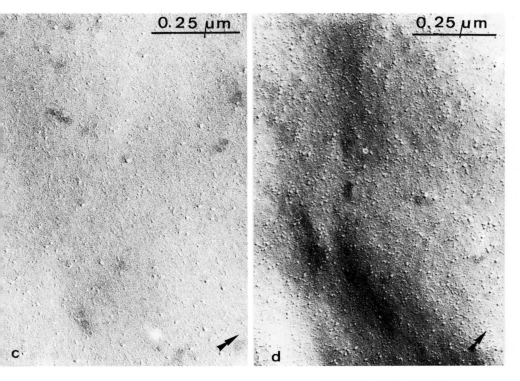

les can be demonstrated in the vicinity of the plasma membrane (→). (c) and (d) split-halves of the outer cell membrane in freeze-fracture. Differences in the numerical density of the membrane particles between the E-fracture face (c) and the P-fracture face (d). Direction of shadowing: →→

terized by degeneration of the ultrastructure but, in particular, the endoplasmic reticulum is dilated by vacuolation. The plasmalemma displays ruptures and close-packed cell membrane fragments are found, mostly of a size seen only under the electron microscope (fig. 4b). The analysis of the plasmalemma of the Wish cells treated for 6 days with mercaptopurine, shows in the freeze-fracture a further increase in the integral membrane particles (fig. 4c) on the P-fracture faces with a mean value of 1518 particles per μm^2, in comparison with the untreated Wish cell. With a value of 1.45, the coefficient of dispersion indicates a statistically significant aggregation of particles on the E-faces in particular (table I).

The Wish cells treated under the same culture conditions with Ney-

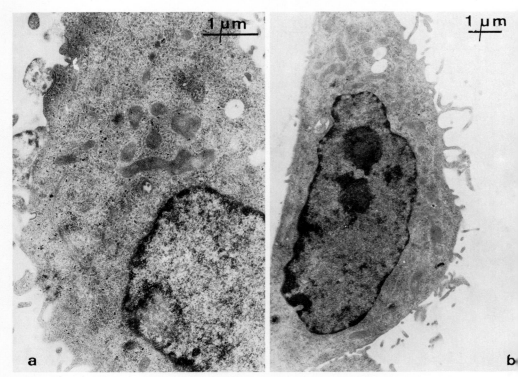

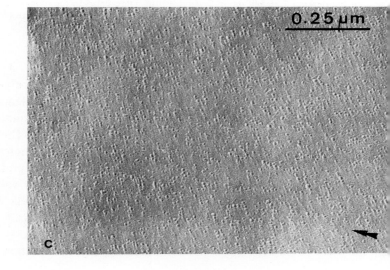

Tumorin are characterized in ultra-thin sections after 10 days by fewer folds of their cell surface than untreated cells, large mitochondria with occasional slight degeneration in the form of vacuoles and large nuclei with finely dispersed chromatin. The cells are often grouped in clumps without evidence of more of the characteristic cell contact points, such as gap-junctions, than were found, by way of comparison, in untreated Wish cell cultures. In the freeze-fracture, the plasma membranes display a reduction in the number of particles on the P-faces (fig. 5c) with a mean value of 975 per μm^2 and a random distribution, as compared with untreated Wish cells (table I).

In cultures treated for 17 days with NeyTumorin, there are again more folds in the cell surface and the nuclei display a roughly dispersed chromatin and prominent nucleoli (figs. 5a and b). Degenerative cell structure changes are absent. In the freeze-fracture, the number of particles on the P-faces of the plasmalemma is again slightly higher as compared with the culture treated for 10 days (fig. 5d) but the mean value does not reach the untreated-culture value (table I).

Density and Topography of Membrane Particles – Morphological Indicators of Tumor Genesis

In summarizing, it can be stated that the results of this pilot study provide a general overview of the possibilities in the assessment of molecular structural changes of cell membranes with the aid of the freeze-fracture method to an extent which was hitherto not possible with other morphological methods.

As originally described, the intramembrane particles in the freeze-fracture represent proteins in the double-lipid layer of the membrane. Little is known about the mechanism by which the topographical distribution of these integral membrane proteins is controlled. Parts of the cytoskeleton of the cell appear to have a major influence on the distribution of

---

Fig. 3, a–c. Human Wish tumor cells in vitro (untreated, 19 culture days). (a) and (b) in ultra-thin section demonstration of increased cell-surface activity with proliferation of cell processes and microvilli. (c) In the freeze-fracture of the outer cell membrane (P-face) clear increase in membrane particles as compared with the control (fig. 2 d). Direction of shadowing: →→

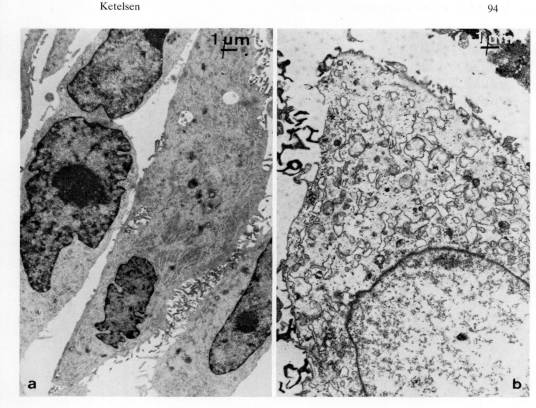
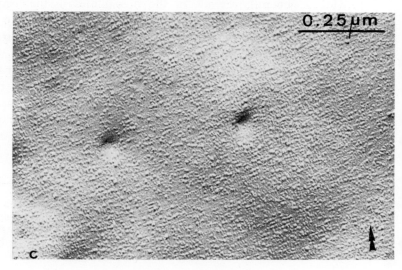

intramembrane particles. Thus, for example, an influence on the cytoskeleton by Cytochalasin B leads to an aggregation of the membrane particles, as demonstrated in vitro by *Speth et al.* [7]. Furthermore, changes in the topographical distribution and numerical density of integral membrane particles can also be associated, however, with a broad spectrum of physiological-metabolic cell conditions [1]. In the neoplastic transformation of the cell, cell membranes and the cell elements associated with them, are obviously subject to various kinds of structural, organizational and functional alterations, so that changes in the numerical density and topography of the membrane particles as morphological indicators can be related to tumor genesis. In our pilot study, this working hypothesis is confirmed by a significant difference in the number of integral membrane particles between normal amnion cells and Wish tumor amnion cells in vitro.

Analogous findings were made as far back as 1976 by *Weinstein* [8] on human urinary bladder epithelium cells in vivo, which were not yet invasive, but malignantly transformed and confirmed in 1978 by *Pauli et al.* in experimental studies on the same organ in the rat [5].

In our pilot study, increase in the integral membrane particles in comparison with the conventional ultrastructural findings at the ultra-thin section correlates with an increasing modification of the cell surface seen in a rise in the number of microvilli formed. This correlation is likewise evident under the treatment with 6-mercaptopurine: there is an increase in both the number of intramembrane particles and in the microvilli formed. In contrast, under the influence of NeyTumorin, there is a reduction in the number of particles with a decline in the proliferation rate after 10 days and relatively fewer microvilli and cell-surface changes are demonstrated than in the untreated Wish cell cultures. After 17 days of treatment with NeyTumorin, there is again a slight rise in the number of intramembrane

---

Fig. 4. Human Wish tumor cells in vitro after treatment with 6-mercaptopurine (daily 1.5×10^{-5} g/2×10^6 cells). (a) Wish cells after 6-day treatment with marked formation of microvilli and cell processes, close-packed cytoplasm, increased rough endoplasmic reticulum and large nuclei with prominent nucleoli; (b) Wish cell after 10-day treatment with turgid nucleus, degeneration of the mitochondria by vacuoles and extensive cisternae of the endoplasmic reticulum; (c) marked increase in the integral membrane particles in the freeze-fracture of the outer cell membrane (P-face) of a Wish cell treated for 6 days. Direction of shadowing: →→

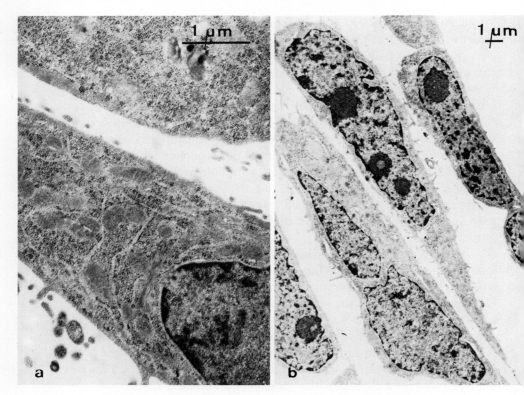

Fig. 5, a–d. Wish cells after treatment with NeyTumorin® (daily 5×10^{-6} g protein/ 2×10^6 cells). (a) and (b) Wish cells after 17-day treatment with well-developed, rough endoplasmic reticulum and mitochondria mostly with dense matrix. Slightly increased cell surface activity (a). The nuclei display a roughly dispersed chromatin and mostly prominent

particles and microvilli in comparison with the 10th day. The proliferation rate had again increased in time at this point.

The influence of NeyTumorin on the Wish cell differed morphologically from that of 6-mercaptopurine and should, therefore, be further investigated also with reference to the dosage question in time-kinetic studies. It should also be checked whether an analogous effect of the preparation can also be morphologically demonstrated in vivo in analogy to the present in-vitro pilot study, e.g., in the animal experimentation model aforementioned of the chemically-induced bladder carcinoma of the rat.

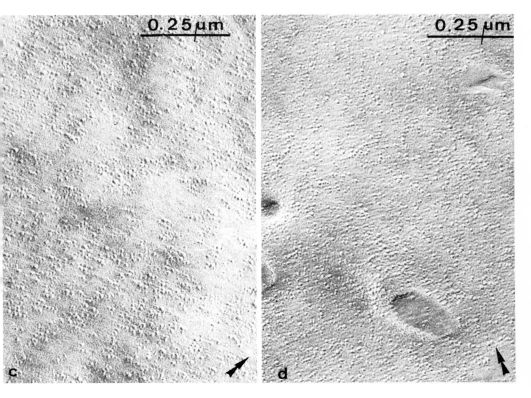

nucleoli (b). (c) P-fracture face of a Wish cell after 10-day treatment. Reduction in the integral membrane-particles in comparison with the untreated Wish cell-culture. (d) P-fracture face of a Wish cell after 17-day treatment. Renewed increase in the number of particles in comparison with the culture treated for 10 days. Direction of shadowing: →→

Summary

The influence of NeyTumorin® on the Wish cell differed morphologically from that of 6-mercaptopurine and should, therefore, be investigated further with reference also to the dosage question in time-kinetic studies. It should also be checked whether an analogous effect of the preparation can be morphologically demonstrated in an in-vivo pilot study, e.g., in the aforementioned animal experimentation model of the chemically-induced rat bladder carcinoma.

Table I. Density and topographical distribution of intramembrane particles of Wish cells in vitro (untreated, cytostatic culture, culture with NeyTumorin®) in comparison with findings from normal amnion cells in vitro

	Amnion cell culture 19 days		Wish untreated 19 days		Wish + 6-mercaptopurine 6 days		Wish + NeyTumorin 10 days		Wish + NeyTumorin 17 days	
	PF	EF	PF	EF	PF	EF	PF	EF	PF	EF
Number of intramembrane particles (IMP per μm^2 + SD)	693.14 ± 102.41	284.5 ± 7.77	1414.72 ± 297.86	407.25 ± 138.58	1518.66 ± 249.80	366.50 ± 166.10	975.42 ± 319.60	–	1020.0 ± 317.76	162.57 ± 39.55
Coefficient of dispersion (C.D.) of the intramembrane particles ± SD	0.755 ± 0.124	1.019 ± 0.46	1.044 ± 0.49	1.422 ± 0.54	1.148 ± 0.51	1.452 ± 0.28	0.75 ± 0.25	–	0.616 ± 0.05	1.235 ± 0.198

SD = standard deviation

Zusammenfassung

Der zum 6-Mercaptopurin morphologisch unterschiedliche Einfluß des NeyTumorin auf die Wish-Zelle sollte deshalb in zeitkinetischen Studien auch im Hinblick auf die Dosisfrage weiter untersucht werden. Darüber hinaus wäre zu überprüfen, ob sich eine analoge Wirkung des Präparates auch in vivo, z. B. in dem erwähnten tierexperimentellen Modell des chemisch induzierten Blasenkarzinoms der Ratte, morphologisch in Analogie zu der vorliegenden in-vitro-Pilotstudie belegen läßt.

References

1 Benedetti, E. L.; Dunia, I.; Olive, J.; Cartaud, J.: Modulation of plasma membrane architecture in animal cells; in Nicolau, Paraf (eds.), Structural and kinetic approach to plasma membrane functions, p. 6076 (Springer, Berlin-Heidelberg-New York 1977).
2 Branton, D.; Deamer, D.: Membrane structure. Protoplasmatologia II/E/1: 1–70 (1972).
3 Ketelsen, U.-P.: Quantitative freeze-fracture studies of human skeletal muscle cell membranes under normal and pathological conditions; in Angelini, Danieli, Fontanari (eds.), Muscular dystrophy research. Advances and new trends. Excerpta Med. Int. Congr. Ser., No. 527, pp. 79–87 (1980).
4 Kocher, O.; Amaudruz, M.; Schindler, A.-M.; Gabbiani, G.: Desmosomes and gap junctions in precarcinomatous and carcinomatous conditions of squamous epithelia. – An electron microscopic and morphometrical study. J. submicrosc. Cytol. *13:* 267–281 (1981).
5 Pauli, B. U.; Friedell, G. H.; Weinstein, R. S.: Topography and numerical densities of intramembrane particles in chemical carcinogen-induced, urinary bladder carcinomas in Fischer rats. Lab. Invest. *39:* 565–573 (1978).
6 Singer, S. J.; Nicolson, G. L.: The fluid mosaic model of the structure of cell membranes. Science *175:* 720–731 (1972).
7 Speth, V.; Bauer, H. C.; Brunner, G.: Changes of the internal organization of the plasma membrane correlated to the regeneration potency of the cell. Biochim. biophys. Acta *649:* 113–120 (1981).
8 Weinstein, R. S.: Changes in plasma membrane structure associated with malignant transformation in human urinary bladder epithelium. Cancer Res. *36:* 251–252 (1976).

Prof. Dr. med. U.-P. Ketelsen, Universitäts-Kinderklinik, Mathildenstraße 1, D-7800 Freiburg i. Brsg. (FRG)

Pharmacological and Toxicological Studies

Influence of NeyTumorin®-Sol and Subfractions on the Growth Behavior of Tumor Cells in Vitro

Th. Stiefel

Stuttgart, FRG

The xenogenic peptide and protein preparation, NeyTumorin®-Sol, is a complex mixture of peptides and proteins with molecular weights ranging from about one thousand to one million. The active substances are isolated by a patented manufacturing process from the following xenogenic organs and tissues: diencephalon, maternal placenta, Funiculus umbilicalis, Thymus juv., Gland. pineal., Testes juv., Gland. suprarenal., thyreoidea, Medulla oss., pulmo, hepar, pancreas, ren, lien, Mucosa intestinal.

Standardization of Natural Substances

To avoid variations in this natural product from batch to batch, these natural substances must also be accurately standardized. Essentially, there are two possibilities for the standardization of peptides and proteins of therapeutic efficacy:
– Analytical characterization with the aid of analytical methods using high-resolution instruments;
– measurement of biological activity in suitable in-vitro systems.
To guarantee an optimal pharmaceutical quality, vitOrgan Company has developed a comprehensive control system which now ensures both a complete analytical characterization of the preparation and a reproducible biological activity of each individual batch. New developments in the high-pressure, liquid chromatography field now permit the separation by

molecular weight of even such complex substance-mixtures as found in NeyTumorin®. An analytical identity-check with the aid of a fingerprint model, enabling a molecular characterization of the active agents, can then be made. Figure 1 shows the diagram of a calibration mixture as obtained in the separation with a 60-cm HPLC column. The separation time is 60 h at a pressure of 16 bars. When the preparation NeyTumorin®-Sol is applied to the HPLC column, a characterization of the molecular weights contained in this preparation can be carried out on the basis of the calibration-function by means of molecular weight markers (fig. 2). Consequently most of the active agents of NeyTumorin® are found in the molecular-weight range of about 10 000. Other fractions have molecular weights of about 4000 and between 80 000 and 300 000. Other methods of characterizing the active-agent composition of NeyTumorin®-Sol are SDS electrophoresis and amino acid determination. Both methods are used by the control departments of the vitOrgan Company, Stuttgart, FRG. The molecular identity of each batch is guarenteed by these modern, analytical procedures.

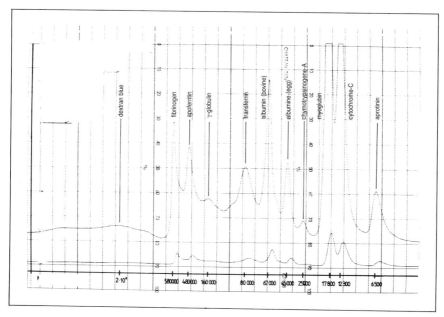

Fig. 1. Absorption diagram at 280 nm of a mol weight standard mixture after separation by a HPLC-column and subsequent evaluation by flow-photometry.

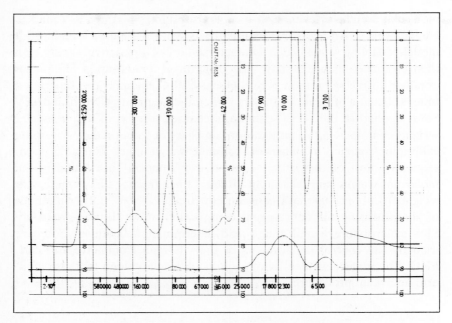

Fig. 2. Absorption diagram of NeyTumorin® at 280 nm after separation by a HPLC-column with subsequent recording by flow-photometry.

Inhibition of Tumor Cells, Stimulation of Normal Cells

In addition to the only descriptive analytical methods, biological activity is, however, a far more important reference parameter with natural substances. An in-vitro test system which supplies maximum information is the human-cell culture of normal and tumor cells. The determination of the thymidine-incorporation rate under incubation with test-substances, permits the assessment of a direct effect of test substances on the proliferation behavior of cells. NeyTumorin®, used in oncological therapy, was thus studied in various human tumor and normal-cell cultures with respect to its effect on the thymidine incorporation rate and, therefore, on the DNA synthesis. As can be seen from figure 1, page 20 (lecture Porcher, this vol.), NeyTumorin® inhibits the DNA synthesis rate of various tumor cells (Wish, HEP II, melanoma), in a dose-related manner as compared with control cultures which were only treated with the solvent. However, diploid fibroblasts tend rather to be stimulated in their DNA synthesis by fairly high concentrations of NeyTumorin® (FH 86). The chemical cytostatic agent, 6-mercaptopurine, likewise tested in these cultures, proved to be cytotoxic, both with tumor and normal cells. The 8-hour incubation of the cells with the isolated thymus components of NeyTumorin®, NeyThymun® k, supplied results comparable to those with NeyTumorin®.

Since, because of various error-sources, assessment of biological activity by thymidine incorporation tests alone is not enough, growth-curves of various cells during incubation with NeyTumorin®-Sol, or subfractions, were additionally recorded. In each case, the effect was compared with controls, which received only the solvent and cultures treated with the purine antagonist 6-mercaptopurine. Figures 3 and 4 show the normal course for tumor cells: Controls treated with physiological saline solution displayed a clear increase in the visually-determined cell counts after an incubation-time of 78 h. After 48 h, NeyTumorin® displays a clear inhibitory effect on the cell proliferation, both in the form of Sol and also as its subfractions. However, in the first 48 h after incubation, there are scarcely any differences to be observed in the cell count between controls treated with physiological saline solution and cultures treated with Ney-

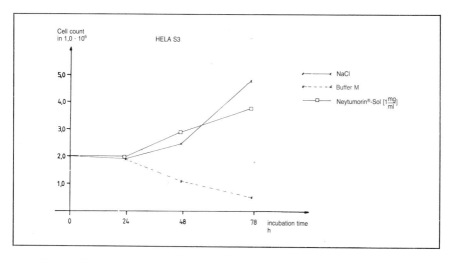

Fig. 3. Growth behavior of HELA-cells under treatment with NeyTumorin®-Sol. 6-mercaptopurine (buffer M) in comparison with physiological saline solution. 5% FBS in the culture medium was added to all cultures. The evaluation was carried out by the visual counting of the cells.

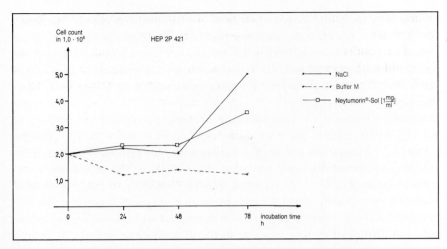

Fig. 4. Growth behavior of HEP-2 cells under the treatment with NeyTumorin®, 6-mercaptopurine (buffer M), and physiological saline solution. The cell counts were determined by visual counting. The cell cultures were given 5% FBS in the medium.

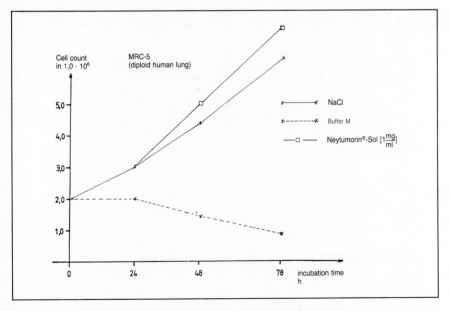

Fig. 5. Growth behavior of diploid normal cells (MRC-5, human lung cells) under incubation with NeyTumorin®, 6-mercaptopurine (buffer M), and physiological saline solution. 5% FBS was added to the culture media. The determination of the cell-count was carried out visually.

Tumorin®. In contrast, the treatment with 6-mercaptopurine is already evident after 24 h in the form of a clear decline in cell count.

The results in normal-cell culture (fig. 5) correspond to the results obtained with respect to the thymidine-incorporation rate: already after 24 h, 6-mercaptopurine displays a cytotoxic action, whereas NeyTumorin®-Sol, when compared with controls treated with physiological saline solution, rather tends to stimulate the cell growth of normal cells.

With the analytical methods described here, and the in-vitro test systems on human tumor and normal cells, quality control methods are available which guarantee both identity and purity in the biological sense, permitting a standardization of the biological activity of NeyTumorin®.

Summary

With the analytical methods described here and in-vitro test systems on human tumor and normal cells, methods are available for quality control, which guarantee both identity and purity in the biological sense, permitting a standardized biological activity of Ney-Tumorin®.

Zusammenfassung

Mit den hier vorgestellten analytischen Methoden sowie den in-vitro-Testsystemen an humanen Tumor- und Normalzellen stehen der Qualitätskontrolle Methoden zur Verfügung, die sowohl die Identität und Reinheit in biochemisch-analytischem Sinne gewährleisten als auch eine Standardisierung der biologischen Aktivität von NeyTumorin® zulassen.

Dr. Th. Stiefel, Olgastr. 139/2, D-7000 Stuttgart 1, FRG

Biological Investigations with NeyTumorin®-Sol and NeyTumorin® E-Sol

G. Gillissen

Dept. of Medical Microbiology, Medical Faculty, Aachen, FRG

Classical therapy of malignant tumors is, in many respects, still unsatisfactory. It is, therefore, quite understandable that it was attempted to achieve therapeutic effects via manipulation of the immune defense mechanisms. In this context, observations on the effect of xenogenic tissue factors are of interest; they have been used in such recent preparations as NeyTumorin®-Sol and NeyTumorin® E-Sol.

Thymus and spleen extracts inhibit DNA synthesis of leukemia thymocytes up to 80% [5]; a growth inhibition of leukemic cells has likewise been reported [10]. Prophylactic doses of bovine decidua extracts led to a reduction of tumor rate after methylcholanthrene application [28]. Using fractions of such extracts, it could be demonstrated that the DNA synthesis of normal cells is stimulated, while that of tumor cells is inhibited [12, 13]; the same was confirmed with cultures of human cells [21]. This effect could be achieved with extracts of different organs; hence, the effect is not organ-specific [20]. The best result was achieved when using a sulphatized preparation (NeyTumorin) from different xenogenic tissues, such as thymus juv., placenta mat., liver, bovine, and pork pancreas [20]. The effect could be demonstrated in a Meth.-A sarcoma, a transplantation tumor in mice, as well as in 3-Lewis lung carcinoma in C57Bl6 mice, and in a L 1210-suspension tumor in DBA mice. The effect, therefore, is also not tumor-specific [20].

The effect may equally be explained by an activation of the immune defense (inefficacy in animal strains without thymus) as well as by a selective inhibition of tumor cells – therefore, it is a direct effect (inhibition of DNA synthesis) [20]. It is assumed that the active material is a relatively low molecular fraction, a protein or polypeptide [19].

NeyTumorin®-Sol and NeyTumorin® E-Sol are relatively new pharmaceutical products, which are administered subcutaneously or intravenously and apparently are tolerated very well. For clinical use, a number of requirements – apart from the therapeutic efficacy – are important; they will be discussed here as far as results are available.

The following parameters have been examined: Sterility, LD 50 of the preparations, as well as pyrogenicity according to international standards. Evidence for therapeutically used substances, not showing mutagenic effects has found increasing interest. Since particularly cellular-immune

Table I. Sterility testing

	European Pharmacopeia, 1st suppl., 1980	USP XX
Material to be tested	Removal under sterile conditions; e.g., Laminar Flow Bank	Idem
Culture media	Type and preparation: Prescribed; others are possible provided a sufficiently high efficacy being demonstrated (media for aerobic and anaerobic germs, also for yeasts and fungi)	Prescribed
	Incubation time: At least 7 days at 30–35° C and 20–25° C, resp.	Same, but up to 14 days
	Efficacy: Germ growth when contaminated with small inocular of defined strains of microorganisms; maximum incubation time = 7 days	Preliminary tests with 2 aerobic and anaerobic species
Test material	Exclusion of antimicrobial activity; to be tested with defined strains (2–4 fold); in case of no or delayed growth: dilution or addition of inactivators	Same with different bacterial strains and fungi
Testing	In case of an aqueous solution: Dilution with medium 1:10; incubation at 30–35° C (bacteria) and 20–25° C (fungi) for 14 days	At least for 7 days
	Quantities (for parenteralia): The whole content of a vial or only 10% depending on the sample quantity. Pooling is done when single vial content is insufficient	Idem
	In case of germ growth: Test to be repeated. If again germ growth but due to different species, a further assay is indicated with twice the number of samples. If no growth: sterile	
	Filter method (enrichment): In case of larger volumes	Idem

defense mechanisms seem to be involved [8, 9, 11, 14, 25, 26] regarding the effects of some proteins on tumors and tumor defense respectively, the influence of these preparations on the so-called "footpad-swelling reaction" was included as an experimental model.

(1) Sterility Testing

As sterility is of major clinical importance, especially for parenterally applied substances, the methods for sterility-testing are prescribed in detail by the different pharmacopeias. Although necessarily similar in principle, the specifications differ slightly from country to country. As an example, the specifications of European and US pharmacopeia (USP) are compared. Table I shows that the testing procedures are more extensively described and defined by the European pharmacopeia than by the USP. Moreover, they are also somewhat stricter in some points, such as the incubation time. The testing of the two preparations was carried out according to the here valid specifications of the European pharmacopeia (for results see table II).

For each substance to be tested, proof must be given that it does not inhibit microbial growth by itself. The importance of this question can be demonstrated by the fact that micro-organisms (e.g., spores) may even survive in ethanol, but may multiply in appropriate dilutions (or, in case of other substances, in the presence of inactivators). It has, therefore, to be demonstrated in preliminary tests that there is no growth inhibition when contaminated artificially with different microbial species.

Table II shows that this is the case for NeyTumorin®-Sol, as well as for NeyTumorin® E-Sol and Staph. aureus. A corresponding dilution factor was, therefore, used for all sterility tests. The preparations proved to be sterile.

(2) Pyrogenicity Testing

For evaluating the presence of pyrogens, the European pharmacopeia allows only the experiments with rabbits. The USP also allows the so-called "limulus-test" (limulus-amoebocytes-lysate test) for the final control of a preparation. This is an in vitro test which is simple and quick and can be carried out quantitatively with very high sensitivity (10–50 pg of

Table II. Sterility tests with NeyTumorin®-Sol and NeyTumorin® E-Sol

(A) Preliminary tests to determine the mean inhibitory concentration (MIC)

Micro-organisms[1]	MIC – mg per ml	
	NeyTumorin®-Sol	NeyTumorin® E-Sol
Staph. aureus ATCC 6538	0.47	>1.88
E. coli ATCC 8739	>1.88	>1.88
Ps. aeruginosa ATCC 9027	>1.88	>1.88
Cand. albicans ATCC 10231	>1.88	>1.88

(B) Results

Media: Soybean-Casein-Digest medium (USP); thioglycolate broth; Sabouraud medium; glucose broth; blood agar; DST agar; McConkey agar; Sabouraud agar and Czapek-Dox medium.

Sterility of media and efficacy when contaminated with small inocula (germs see above) established.

Parameters	NeyTumorinF7-Sol	*NeyTumorin*® E-Sol
Solvent	Original solvent	Saline
Initial concentration	7.5 mg/ml	7.5 mg/ml
Concentration in the test system	0.15 mg/ml	0.38 mg/ml
Limit of evidence (colony-forming units (CFU) per mg)	1 CFU/7.5 mg	4 CFU/7.5 mg
After 14 days		
at 20–25° C	no growth	no growth
at 30–35°C	no growth	no growth
Subcultures on solid medium	no growth	no growth

[1] Inoculation = 1×10^5 microscopically adjusted cells per ml culture medium. Media: Soybean-Casein-Digest medium and Sabouraud medium.

coli-endotoxin/ml) [27]. However, there are known interfering factors which may lead to false positive, but also to false negative results and which are to be accounted for as far as known for a given preparation. In contrast, using rabbits, a rise in temperature is tested directly in vivo. A preparation has, therefore, to be excluded, also when a rise in temperature is not due to lipopolysaccharides, which is reasonable from a clinical point of view.

The differences between the European pharmacopeia and the USP (table III) consist only in the possibility of a repeated use of animals for

Table III. Pyrogenicity Testing

	European Pharmacopeia, 1st suppl., 1980	USP XX
Type of test	Tests only with rabbits	Tests with rabbits; Limulus test also applicable
Animals	Male or female rabbits of at least 1.5 kg. Feeding without addition of antibiotics. No weight loss 1 week before test. During the night before testing: no feeding, only water.	Idem
	Animals cannot be used if (1) they have been used some time before in a positive test with rise in temperature of $> 1.2°$ C; (2) the animal gave a positive test within 3 weeks prior to testing; (3) the animal has been used in a (negative) test within 3 days prior to testing; animals with initial temperature $> 39.8°$ C or $< 38.0°$ C.	Repeated use after intervals of at least 48 h in case of negative tests; intervals of at least 2 weeks after positive tests
Testing	Keeping: Quiet room; testroom within the same temperature range as room where animals are kept. Instruments like glassware and syringes = pyrogen-free.	Idem
	Animals in sitting position with loose neck support. Preliminary test: 1–3 days before testing = 10 ml of pyrogen-free sterile saline are slowly injected intravenously.	Idem Preliminary test only before first use
	Test solution pre-warmed to 38.5° C over 4 min, injected intravenously (> 0.5 ml – < 10.0 ml/kg). Measuring time: 90 min before to 3 h after injection of substances to be tested. Initial temperature: $\bar{x}$ of 2 measurements with an interval of 30 min before test (not to be used if difference $> 0.2°$ C). Maximum value: Highest temperature within 3 h after injection.	
	Evaluation: The sum of maximum deviations must be smaller than a tabulated value dependent on the number of animals; no single deviation of $> 0.6°$ C. Repeat if the sum is higher than tabulated value, but lower than the maximum value in table.	Idem Repeat with 5 animals and evaluate sum of maximum deviations of all animals

tests and in the frequency of preliminary tests. Also, the European pharmacopeia is somewhat stricter.

The tests carried out according to these specifications showed that the preparations NeyTumorin®-Sol and NeyTumorin® E-Sol are clearly free of pyrogens (table IV).

(3) The Lethal Dose 50% (LD 50)

The LD 50 is the dose per kg at which 50% of the animals die. It is usually calculated according to the Chi^2-method [15]. Acute and chronic toxicity have to be differentiated. Whereas in the former case, a given dose is administered once, there are several applications in the latter. It will necessarily always be a point for discussion in just how far results of animal experiments can be valid for humans; particularly since reactions of different animal species – even of different strains of the same species may vary considerably. Moreover, the application method is equally important, i.e., whether given orally, intramuscularly, subcutaneously or intravenously. The spectrum of possible kinds of tests is, of necessity, very

Table IV. Pyrogenicity Testing in Rabbits

Animal No.	Initial temperature			Max. deviation	Max. temperature	$\Delta° C$
	t_1	t_2	$\bar{x}$			
NeyTumorin®-Sol[1]						
1	38.80	38.70	38.75	0.10	39.30	0.55
2	38.30	38.20	38.25	0.10	38.40	0.15
3	38.40	38.20	38.30	0.20	38.40	0.10
						$\Sigma = 0.80$
NeyTumorin® E-Sol[2]						
4	38.40	38.40	38.40	0	38.60	0.20
5	38.20	38.30	38.35	0.10	38.40	0.05
6	38.20	38.20	38.20	0	38.65	0.45
						$\Sigma = 0.70$

[1] Dose per animal = 1.5 ml of original solvent containing 11.25 mg of the preparation
[2] Dose per animal = 1.5 ml of pyrogen-free saline, containing 11.25 mg of the preparation

large. Experiments made so far with NeyTumorin®-Sol and NeyTumorin® E-Sol are herewith briefly summarized:

Both preparations are obviously well-tolerated as far as can be concluded from the experience available so far.

In animals – mostly mice of different strains – NeyTumorin-Sol was generally administered subcutaneously in doses of approximately 1 mg per animal [19, 20]. This corresponds to 50 mg/kg or – for a patient of 70 kg – to a dose of 3.5 g.

In patients, NeyTumorin®-Sol was given intravenously or intramuscularly in doses of 30 mg (= 2 vials) corresponding approximately to 0.4 mg/kg [4] or of 300 mg (= 20 vials) by infusion in 1000 mg of physiological saline corresponding to approximately 4.3 mg/kg [24]. In all cases, treatment was well tolerated.

For evaluation of the LD 50 in our own experiments, male Balb/c ABOM mice of 20 ± 1 g (Bomholtgård, Ry, Denmark) were used. The doses of NeyTumorin®-Sol and E-Sol respectively, were administered intravenously, diluted in 0.5-ml saline.

General observations:

(a) NeyTumorin® E-Sol is appreciably better tolerated than NeyTumorin®-Sol.

(b) Tolerance for NeyTumorin®-Sol is considerably increased when the substance is injected slowly.

In detail, the following can be said: For the animal species and strain used, the LD 50 for NeyTumorin®-Sol was 40 mg/kg, with a confidence limit of 32.7–49.0 mg/kg. It is to be considered that the dose per animal is diluted in a volume of 0.5 ml of saline and administered intravenously. It can be assumed that after subcutaneous application and hence a slower resorption, the tolerance would even be better. This is also evident from the fact that the values mentioned could only be achieved by a very slow injection of over 1–2 min.

For NeyTumorin® E-Sol, reliable data for the LD 50 could not be obtained. This was mainly due to the fact that for the same test model the tolerance limit was about 10 times higher and sufficient material of this new preparation was not available. However, the estimated LD 50 for NeyTumorin® E-Sol is approximately 450 mg/kg, as 400 mg/kg were well tolerated (n = 3); at 450 mg/kg, 50% (n = 8) and at 500 mg/kg (n = 8), 37.5% of the animals survived.

In summarizing, it can be said that the tolerance for both preparations, but particularly for NeyTumorin E-Sol, is very high.

(4) Mutagenicity Testing

Therapeutically used substances are increasingly tested for possible mutagenicity. This is important because mutagens can also be considered to be cancerogens. On the other hand, about 85% of cancerogens are also mutagens [16]. Mutagenicity-testing in animals is highly expensive as far as number of animals, time and costs are concerned. Therefore, an in vitro test, the so-called *Ames* test [1], has been developed for screening. In this test system, histidine auxotrophous mutants of salmonella typhimurium are used. Mutagenicity of a substance is determined by the number of revertants exceeding spontaneous re-mutation, i.e., by the induction of a re-mutation from histidine requirement to prototrophy. Mostly, a total of 5 strains is used, all possessing the characteristic of histidine auxotrophy, but with different genetic determinations. This implies that not all strains have to be affected in the same way by a given mutagen [1]. The following strains are used: TA 1535, TA 1537, TA 1538, TA 98, TA 100. The two latter strains have been obtained from TA 1535 and TA 1538 by insertion of a R-plasmid [17]. The strain T 100 is the most sensitive one. It also shows, however, the highest rate of spontaneous revertants. In addition to this test using salmonella typhimurium strains, other in vitro microbial systems have been developed. The *Ames* test is, however, the one most widely used.

The significance of this test method has, of course, a crucial point. It is assumed that the *Ames* test may yield up to 30% false positive results [7]. On the other hand, a negative in vitro test does not necessarily guarantee a negative result in other test systems using animals [3]. However, of all substances tested, obviously all important and recognized cancerogens gave a positive *Ames* test.

Apart from other microbiological details, the use of a liver homogenate, adjusted for its biological activity or the microsome fraction (S 9) of rats respectively, is also crucial for the significance of results obtained by the *Ames* test [2, 6, 23], the rats having been pre-treated with Aroclor 1254 for metabolic activation of cancerogens.

For evaluation of the *Ames* test, an increase of revertants in the presence of a test substance to more than twice the number of spontaneous revertants is considered as being significant [23]. However, a three- to fourfold increase has also been given as criterion for evaluation [1]. Known mutagens are frequently used as positive controls. For the experiments described here (table V), Aflatoxin B1 (Roth, Karlsruhe, FRG)

Table V. Mutagenicity testing using the Ames test

Substances	Dose/plate mg	Tester strains				
		TA 1535	TA 1537	TA 1538	TA 98	TA 100
Control	–	21.8 ± 2.2 (1.3 ± 0.1)	14.8 ± 3.0 (7.5 ± 1.2)	36.3 ± 3.8 (14.7 ± 1.5)	40.0 ± 6.7 (18.1 ± 3.0)	369.5 ± 14.2 (62.3 ± 2.4)
Solvent for NeyTumorin F7-Sol	–	19.8 ± 1.9 (1.2 ± 0.1)	16.3 ± 2.9 (8.2 ± 1.5)	48.3 ± 6.8 (19.6 ± 2.8)	29.8 ± 3.9 (13.5 ± 1.8)	347.3 ± 38.3 (58.6 ± 6.5)
NeyTumorin®-Sol	1.5	33.0 ± 9.5 (1.98 ± 0.6)	19.8 ± 6.1 (10.0 ± 3.1)	–	36.5 ± 5.5 (16.5 ± 2.5)	417.3 ± 30.3 (70.4 ± 5.1)
	0.75	23.8 ± 7.5 (1.4 ± 0.5)	14.8 ± 3.6 (7.8 ± 1.8)	46.5 ± 14.4 (18.8 ± 5.8)	42.3 ± 4.6 (19.1 ± 2.1)	357.8 ± 32.8 (60.3 ± 5.5)
NeyTumorin® E-Sol	1.5	24.5 ± 15.6 (1.5 ± 0.9)	12.8 ± 2.5 (6.5 ± 1.3)	52.0 ± 34.8 (21.1 ± 14.1)	41.8 ± 5.4 (18.9 ± 2.4)	535.5 ± 35.6 (90.3 ± 6.0)
	0.75	29.8 ± 4.0 (1.8 ± 0.2)	18.8 ± 2.5 (9.5 ± 1.3)	51.5 ± 6.4 (20.9 ± 2.6)	38.5 ± 10.1 (17.4 ± 4.6)	401.5 ± 40.7 (67.7 ± 6.9)
DMSO	–	–	–	51.0 ± 9.1 (20.6 ± 3.4)	24.0 ± 1.4 (10.9 ± 0.6)	296.0 ± 52.5 (49.9 ± 8.9)
Aflatoxin B_1	8 µg	–	–	182.0 ± 33.2 (73.7 ± 13.0)	527.5 ± 34.9 (238.8 ± 15.8)	996.0 ± 138.6 (168.0 ± 23.4)

Mean number of revertants per plate (4 assays in parallel). ( ) = mean number of revertants per 10^6 germs. DMSO (dimethylsulfoxide), solvent for aflatoxin B_1 and control for aflatoxin test. The test solutions did not show any growth inhibition in the concentrations indicated. NeyTumorin®-Sol was diluted in the original solvent and NeyTumorin® E-Sol in saline.

was used, causing a definite increase of the number of revertants only in TA 98 and TA 100. In contrast to other references [17], an effect could also be observed with TA 1538. The tests were carried out as indicated by *Ames et al.* [2, 6].

Table V shows that neither NeyTumorin®-Sol nor NeyTumorin® E-Sol had any mitogenic effect in the *Ames* test. Highest possible concentrations were used for the tests, being in the mg-range for each test plate. This is important as the sensitivity of the *Ames* test lies in the µg or even ng range [7].

(5) Effect on Cellular Immunity

For any immunological influence on tumor growth, cellular mechanisms play a decisive role. In order to study the influence of NeyTumorin®-Sol and E-Sol in this respect, the "footpad-swelling test" was taken as a model [18] using male mice of 20 ± 1 g of the inbred-line Balb/c ABOM.

The principle consists of immunization of mice with sheep erythrocytes, triggering the reaction by subplantar injection of the same antigen. The increase of skin thickness (Δ mm) assessed 24 h after the challenge injection, was taken as a value for the reaction. In one series, animals were treated with NeyTumorin® preparations 1 day and in another series of experiments, 1 week before immunization, because it was shown [2] that a longer interval between treatment and immunization leads to better results.

Tumors may induce a depression of immune response mechanisms. From a clinical point of view it is, therefore, of interest whether a modulation of immune defense can be achieved not only in case of a functionally intact immune system, but also in case of a state of immune depression. Experiments were, therefore, carried out using, apart from normal animals, also animals pretreated with cortisone.

Animals were treated by intravenous injections of the test substances in 0.5 ml saline (NeyTumorin®-Sol = 0.3 mg, and NeyTumorin® E-Sol = 4.0 mg). Cortisone (Hydrocortison, Hoechst, crystalline suspension) was administered subcutaneously in doses of 1.0 mg per 0.5 ml saline 1 day before immunization (6 animals per group).

The experiments showed the following results (table VI):
(1) Both, NeyTumorin®-Sol and NeyTumorin® E-Sol stimulate cellu-

Table VI. The Influence of NeyTumorin®-Sol and NeyTumorin® E-Sol on the "footpad-swelling reaction"

	Without cortisone		With cortisone	
	Δ mm ± s.d.	p	Δ mm ± s.d.	p
Control	0.38 ± 0.01	–	0.06 ± 0.008	–
NeyTumorin®-Sol				
day −1	0.50 ± 0.01[1]	< 0.01	0.10 ± 0.008[3]	< 0.01
day −7	0.55 ± 0.009[1]	< 0.01	0.10 ± 0.01[3]	< 0.01
NeyTumorin® E-Sol				
day −1	0.57 ± 0.004[2]	< 0.01	0.20 ± 0.006[4]	< 0.05
day −7	0.59 ± 0.009[2]	< 0.01	0.21 ± 0.006[4]	< 0.05

Mutual. comparisons:
[1] $p < 0.01$; [2] $p < 0.01$; [3] $p > 0.05$; [4] $p < 0.05$.

lar defense reactions significantly when the preparations are administered as single doses one day before immunization.

NeyTumorin® E-Sol is significantly more efficient than NeyTumorin®-Sol; the former, however, was administered in a higher dose because of its better tolerance.

(2) When the preparations are administered earlier, i.e., 7 days before immunization, but as single doses, the effect is even better than with later application. This result is compatible with other observations of this kind [20].

(3) The "footpad-swelling reaction" is strongly inhibited by cortisone. NeyTumorin®-Sol and NeyTumorin® E-Sol reduce the inhibitory effect of cortisone significantly when administered 1 day before immunization; however, a complete suppression of the cortisone effect could not be achieved.

Also in this case, NeyTumorin® E-Sol is more efficient, although in higher doses.

(4) When the two preparations are administered earlier, i.e., 7 days before immunization – and equally under simultaneous cortisone therapy – there is a small but significant improvement of the effect of NeyTumorin® E-Sol, when compared to later administration (only 1 day before immunization). Under the same experimental condition, NeyTumorin®-Sol does not show this effect, depending on the time of application relative to immunization. On the other hand, this preparation could only be given in smaller doses.

Summary

The effect of NeyTumorin®-Sol and NeyTumorin® E-Sol was tested with respect to different pharmacologically and immunobiologically important parameters. The experiments showed that both preparations were sterile and free of pyrogen, meeting the standards of European and US pharmacopeias. The LD 50 of NeyTumorin®-Sol was 40 mg/kg when using Balb/c mice and under slow intravenous application. For NeyTumorin® E-Sol, the LD 50 of NeyTumorin® E-Sol is supposed to be approximately 450 mg/kg under identical experimental conditions.

Both preparations are not mutagenous in the *Ames* test.

In the "footpad-swelling test", both preparations stimulate the reaction; NeyTumorin® E-Sol (although in higher doses), more than NeyTumorin®-Sol. The stimulatory effect of NeyTumorin® E-Sol was even more marked when the preparation was administered one week prior to immunization.

The inhibitory effect of cortisone on the "footpad-swelling reaction" was significantly reduced, but not completely suppressed by both preparations.

Zusammenfassung

An verschiedenen Versuchsmodellen wurde der immunmodulatorische Effekt von NeyTumorin®-Sol beschrieben. Von Interesse war deswegen die Frage, ob und in welchem Ausmaß dieser Effekt sich auch auf einen Infektionsablauf auswirkt. Mit der experimentellen Infektion unter Verwendung von Staph. aureus konnte gezeigt werden, daß durch eine Vorbehandlung der Versuchstiere mit NeyTumorin®-Sol die Überlebensrate verbessert und darüber hinaus der therapeutische Penicillin-Effekt deutlich verstärkt wird.

Maligne Tumoren können zu einer Immundepression mit der Folge einer erhöhten Infektionsbereitschaft führen. Da andererseits für einen optimalen Effekt einer antimikrobiellen Chemotherapie eine funktionelle intakte Immunabwehr von Bedeutung ist, stellen Infekte bei malignen Tumoren oft ein besonderes therapeutisches Problem dar. Untersucht wurde deswegen der Einfluß von NeyTumorin®-Sol auf eine experimentelle Infektion mit Candida albicans zu Beginn eines palpablen, durch Methylcholanthren induzierten Tumors. Die Versuche ergaben bei Tumortieren gegenüber normalen Tieren eine etwas beschleunigte Mortalitätsrate. Diese wurde durch Gaben von NeyTumorin®-Sol in signifikanter Weise reduziert, was im Sinne einer Verbesserung der Immunabwehr gedeutet wurde.

References

1   Ames, B. N.; Durston, W. E.; Yamasaki, E.; Lee, F. D.: Carcinogens are Mutagens: A Simple Test System Combining Liver Homogenates for Activation and Bacteria for Detection. Proc. natn. Acad. Sci. USA *70:* 2281 (1973).
2   Ames, B. N.; McCann, J.; Yamasaki, E.: Methods for Detecting Carcinogens and Mutagens with the Salmonella/Mammalian-Microsome Mutagenicity Test. Mutat. Res. *31:* 347 (1975).
3   Bingham, E. et al. (Interagency Regulatory Liaison Group): Scientific Basis for Identification of Potential Carcinogens and Estimation of Risk. J. natn. Cancer Inst. *63:* 241 (1979).

4   Douwes, F. R.: Zur Problematik von "Dose-finding-studies" bei biologischen "response modifiern" in der Onkotherapie. Therapiewoche 33: 79 (1983).
5   Ebbesen, P.; Olsson, L.: Stimulatory Effect on DNA Synthesis of Thymus and Spleen Extract from Leukemic AKR Mice. J. Cancer Res. clin. Oncol. 100: 105 (1981).
6   Frantz, C. N.; Malling, H. V.: The Quantitative Microsomal Mutagenesis Assay Method. Mutat. Res. 31: 365 (1975).
7   Gericke, D.: Zur Problematik der mikrobiellen Mutagenitätstests. Onkologie 5: 30 (1982).
8   Goldstein, P.; Luciani, M. F.; Wagner, H.; Röllinghof, M.: Mouse T Cell-Mediated Cytolysis Specifically Triggered by Cytophilic Xenogeneic Serum Determinants: A Caveat for the Interpretation of Experiments Done under "Syngeneic" Conditions. J. Immun. 121: 2533 (1978).
9   Goldstein, P.; Rubin, B.; Denizot, F.; Luciani, M. F.: Xenoserum-Induced Cytolytic "T" Cells: Polyclonal Specificity with an Apparent "Anti-Self" Component, and Cooperative Induction. Immunbiol. 156: 121 (1979).
10 Hall, V.; Wolcott, M.: Modulation of Tumor Cell Growth by Thymus Extracts. Fed. Proc. 40: 3351 (1981).
11 Kedar, E.; Schwartzbach, M.: Further Characterization of Suppressor Lymphocytes Induced by Fetal Calf Serum in Murine Lymphoid Cell Cultures: Comparison with in vitro Generated Cytotoxic Lymphocytes. Cell. Immunol. 43: 326 (1979).
12 Letnansky, K.: Stoffwechselregulatoren der Plazenta und ihre Wirkung in Normal- und Tumorzellen. Exp. Pathol. 8: 205 (1973).
13 Letnansky, K.: Tumorspezifische Faktoren der Plazenta und Zellproliferation. Exp. Pathol. 9: 354 (1974).
14 Levy, R. B.; Shearer, G. M.; Kim, K. J.; Asofky, R. M.: Xenogenicserum-Induces Murine Cytotoxic Cells. Cell. Immunol. 48: 276 (1979).
15 Litchfield, I. T.; Wilcoxon, F.: A Simplified Method of Evaluating Dose-Effect Experiments. J. Pharmac. exp. Ther. 96 (2): 99 (1949).
16 McCann, J.; Ames, B. N.: A Simple Method for Detecting Environmental Carcinogens as Mutagens. Am. N.Y. Acad. Sci. 271: 5 (1976).
17 McCann, J.; Spingarn, N. E.; Kobori, J.; Ames, B. N.: Detection of Carcinogens as Mutagens: Bacterial Tester Strains with R Factor Plasmids. Proc. natn. Acad. Sci. USA 72: 979 (1975).
18 Miller, T. E.; Mackaness, G. B.; Lagrange, P. H.: Immunopotentiation with BCG. II. Modulation of the Response to Sheep Red Blood Cells. J. natn. Canc. Inst. 51: 1669 (1973).
19 Munder, P. G.: Experimentelle Untersuchungen über den antitumoralen Wirkungsmechanismus von NeyTumorin®. Therapiewoche 33: 71 (1983).
20 Munder, P. G.; Stiefel, Th.; Widmann, K. H.; Theurer, K.: Antitumorale Wirkung xenogener Substanzen in vivo und in vitro. Onkologie 5: 1 (1982).
21 Paffenholz, V.; Theurer, K.: Einfluß von makromolekularen Organsubstanzen auf menschliche Zellen in vitro. I. Diploide Kulturen: Z. Kassenarzt 27: 5218 (1978); II. Tumorzellkulturen: Z. Kassenarzt 19: 1876 (1979).
22 Seyfarth, H.: HGMF – eine neue Methode zur Keimzahlbestimmung. Pharm. Ind. 44: 821 (1982).
23 Skopek, Th. R.; Liber, H. L.; Kaden, D. A.; Thilly, W. G.: Relative Sensitivities of Forward and Reverse Mutation Assays in Salmonella Typhimurium. Proc. natn. Acad. Sci. USA 75: 4465 (1978).
24 Stiefel, Th.: Intravenös-gängiges NeyTumorin®-Sol. Therapiewoche 33: 74 (1983).
25 Tsutsui, J.; Everett, N. B.: Specific Versus Non-Specific Target Cell Destruction by T-Lymphocytes Sensitized in vitro. Cell. Immunol. 10: 359 (1974).
26 Watson, J.; Gillis, S.; Marbrook, J.; Mochizuki, D.; Smith, K. A.: Biochemical and

Biological Characterization of Lymphocyte Regulatory Molecules. J. Exp. Med. *150:* 849 (1979).
27 Watson, St. W.; Levin, J.; Novitsky, Th. J.: In Endotoxins and their Detection with the Limulus Amebocyte Lysate Test. Progr. in Clin. and Biolog. Research, vol. 93 (Alan R. Liss, Inc., New York 1982).
28 Wrba, H.: Krebsverhütung und Verhinderung der Krebsentstehung. Österr. Ärztezeitung *29:* 1351 (1974).

Prof. Dr. Dr. G. Gillissen, Institut für Mikrobiologie der RWTH Aachen, Pauwelsstrasse, D-5100 Aachen (FRG)

Flow Cytometry and Surface-Marker Phenotyping Using Monoclonal Antibodies: A Combined Approach to Precisely Define the State of the Immune System

Martin R. Hadam

Oncological Laboratory, Dept. of Children's Surgery, Children's Hospital, Medical University Hannover, Hannover (FRG)

Introduction

Immunological characterization of different lymphocyte populations in its early days, relied on the demonstration of surface immunoglobulins for B-lymphocytes and on the rosetting phenomenon with sheep erythrocytes to define T-lymphocytes. Other surface markers which could be used to distinguish cell populations were Fc-receptors and complement receptors; again both by rosetting-procedures, using the respective ligands of the thus-defined receptors. Production of xeno-antisera against subsets of cells proved to be most difficult due to the low titers obtained and the inherent problems and inconsistencies of the absorption process. Moreover, supply was limited and barely reproducible. Alloantisera from certain patients were shown to be subset-specific; however, limited access did not support their widespread use.

With the advent of the monoclonal antibody technique by Köhler and Milstein, new tools for the dissection of heterogenous cell populations were provided. Thus, antibodies could be tailored to the specific needs of the researcher; these were of unsurpassed specificity, without any absorption, and provided an unlimited supply. However, specificity had to be carefully evaluated in each case to control unexpected cross-reactions. Since then, a vast number of monoclonal antibodies against human (and murine) leukocyte differentiation antigens has been produced. Since most of the relevant reagents are commercially available, the immunological discrimination of cellular subsets is now widely applicable. This phenotyping approach is totally different from the methods previously used and has – since its introduction in the late seventies – already generated a new view toward the immune system.

Automated quantification of cellular parameters, from cells in suspension has been a major goal of hematologic technology for a long time. The early instruments produced allowed measurements of cell volume, now widely known as the Coulter counter. Additional lines of development concentrated on cellular fluorescence, in particular on the distribution of fluorescently labeled DNA and RNA to quantitate cellular aneuploidy, assisting in cancer diagnosis. Cell-surface-associated fluorescence was also measured, using appropriate antisera. The rapid development of microelectronics later provided instruments which were easier to handle, while extending their capabilities. When these technological improvements merged with the achievements of producing monoclonal antibodies, both methods proved to be mutually stimulating. At present, they provide us with the most elaborate technological tool, allowing the sophisticated use of the novel reagents available.

The following presentation is aimed toward those not acquainted with these recent technological developments. Hopefully, it will provide basic information on how to properly interpret flow-cytometric data presentations and introduce the novice to some of its basic applications. It will consist of two parts: First, a general description of a flow-cytometric instrument, its basic features and capabilities and a notion on the labeling procedure used for phenotypic analysis. The second part consists of representative examples from the author's experience, in order to demonstrate the various operating modes of the instrument and some useful applications of the phenotyping approach.

Technical and Methodological Aspects

Most units used for flow cytometry have principally the same setup depicted in figure 1. Thus, every flow instrument comprises three main parts:
- Fluidics system used for sample delivery;
- optical system with light sources and suitable detector systems;
- a computer for immediate data analysis and storage.

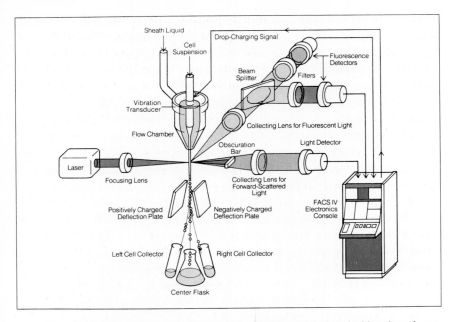

Fig. 1. Schematical view of a flow cytometer with flow chamber, incident laser beam, scatter and fluorescence optics and electronics console. Vibration transducer and deflection plates are for preparative sorting of cells only (reproduced with kind permission of BD-FACS systems).

When a flow cytometer is additionally equipped with a device for sorting cells, the fluidics system will contain a device for breaking up the liquid jet into small droplets which is usually accomplished by means of a piezo element on top of the flow chamber.

The purpose of the fluidics system is to generate from a given particle, or cell suspension, a stream of single cells which are linearly aligned and may then pass sequentially on to the detector system. This is achieved in the flow chamber by a technique called "hydrodynamic focussing", whereby the sample is injected in the center of a laminar jet of sheath fluid. The resulting acceleration at the orifice will force the cells to travel downstream, one after another. The intersection point with the optical system may either directly meet the liquid jet (stream in air system) or may be contained within a cuvette, showing different performance with regard to some parameters and impeding preparative sorting.

Lasers are usually selected as light source, due to their constant, high energy output; however, other sources such as mercury arc lamps are useful, provided fluctuations in light emission are controlled. As can be seen from figure 1, each cell passing the flow chamber and entering the liquid jet will interact with the illuminating laser beam in several ways. Any particle passing the focussed laser beam will scatter the incident laser light. This (forward) scattered light is collected at a small angle to the incident beam and resulting light pulses are measured on a photomultiplier. The resulting signals are proportional to the cross-sectional area of the passing particle or cell. Thus, simply by analyzing the (forward) light-scatter signals of a cell population, one may derive a distribution of the cell size of the cell population under study. It should be noted that, for this type of analysis, no fluorescent labeling is required.

However, if the cells under study are fluorescently labeled, the interaction of a labeled cell with the exciting laser beam will have an additional effect: If the exciting laser wavelength matches the excitation spectrum of the fluorescent dye used, this labeled cell will then emit a fluorescent light pulse of the appropriate wavelength. To measure these fluorescent light pulses, an optical detector system is focussed to the intersection of the laser beam with the liquid jet. To analyze light pulses of differing wavelength, i.e., color, dichroic mirrors (beam splitter) and appropriate filters are included to the light path leading to the fluorescence detectors. Thus, in addition to the scatter signal generated from any particle, fluorescently-labeled cells will produce electrical pulses at the respective photomultiplier detectors, which are proportional to the intensity of the emitted light.

Moreover, since the fluorescence detector system is usually mounted at $90°$ angle to the incident laser beam, another type of light-scatter signal may be additionally obtained. Since these signals are very strong, they are easily recovered by an additional beam-splitter made of quarz (not shown in fig. 1). Measuring pulses of scattered light at $90°$ (right-angle scatter) will provide information on the internal structure of the passing cell such as cytoplasmic granules, nuclear/cytoplasmic ratio and nuclear structure. This right-angle scatter signal is again independent of the fluorescent label and provides an elegant means of differentiating complex cell populations, as will be seen below. Altogether, a fluorescently labeled cell, while passing the exciting laser beam, may produce up to four independent, coincident pulses at the different detectors:

(1) A forward light-scatter signal, related to cell size;

(2) a right-angle light-scatter signal, related to structuredness of the cell;

(3) a fluorescence signal of fluorochrome A (e.g., fluoresceine), and
(4) a fluorescence signal of fluorochrome B (e.g., phycoerythrin).

These four concurrent signals must be handled by the instruments' computer, as shown in figure 2. Here the four light pulses from a single cell have already been transformed by the photomultipliers into electrical pulses which are denoted 1, 2, 3, 4. To reduce electronical noise, one signal (usually forward light-scatter), is selected to trigger a coincidence circuit allowing only coincident pulses to be further processed. Since the height of the electrical pulses is proportional to the intensity of the inducing light pulses, they will first enter an analog-to-digital converter (ADC) which transforms the peak height into a numerical value, which ranges in flow cytometry usually between 1 and 256 so-called channels (other commonly used values are 512 and 1024 and 64 × 64 in correlated matrices). Light pulses of the same intensity will be localized to the same channels and in time will pile up to produce a histogram. Thus, one might generate four independent histograms from a given cell population (i.e., one for each parameter measured).

However, since these signals are correlated in time, they may also be used to describe correlations between the different parameters. In order to obtain intelligable distributions, only two parameters may be correlated at a time. In our example in figure 2, parameter 1 and 2 are chosen to produce a matrix where any point inside will correspond to a defined

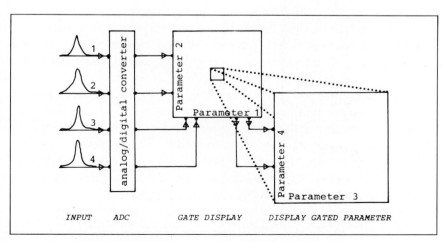

Fig. 2. Principle of multiparameter analysis and gating.

correlation between the two parameters. In practice, this could be forward versus right-angle light-scatter. The remaining two parameters (3 and 4; e.g., two different fluorochromes), are then investigated after a process called "gating": This means that only those pulses 3 and 4 are displayed on another matrix, whose corresponding values of parameter 1 and 2 fall within a certain predefined limit, the so-called "gate". Thus, gating allows the selective analysis of differing cell populations (e.g., defined by forward versus right-angle light-scatter) within a heterogenous mixture. When all incoming pulses are stored as correlated raw data, without any gating (a process which is called "list mode"), gate settings may be defined later while rerunning the data through a computer, and multiple gates can be established from a single experiment, thus providing additional information out of a given experiment. Since gating is such a powerful means of manipulating flow-cytometric data, its use must be indicated at every occasion.

Standard flow cytometers usually comprise one argon laser as a light source. However, for optimal two-color fluorescence work, a second laser

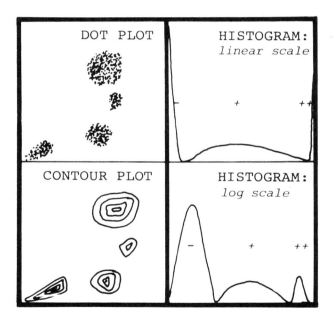

Fig. 3. Data presentation: Left panels are displays of correlated data (e.g., forward vs. right-angle scatter). Right panels are histograms with differing fluorescence scales at the abscissa (linear vs. logarithmic), ordinate being cell number.

(presently another argon laser coupled to a dye laser seems most appropriate) is aligned so that its intersection point with the liquid jet is located some micrometers downstream, compared to the first laser (not shown in fig. 1). Fluorochromes illuminated by this second laser beam produce light pulses which are delayed for the time required by the cell to travel that distance. A special electronic delay circuit will allow only the delayed pulses to pass the coincidence circuit. The major advantage of two-laser flow cytometry resides in its indifference to overlapping fluorescence emission spectra, provided fluorochromes can be differentially excited. Since the instruments' setup is usually more complicated, compared to single-laser work, other solutions are indicated for routine purposes. However, it is recommended for sophisticated applications and – more importantly – an additional photomultiplier allows analyzing concurrently three different fluorescent labels, in addition to the two scatter signals.

The results of a flow-cytometric experiment may be reported in a number of ways, which shall be discussed briefly (fig. 3). Basically, there are two differing output formats: One may either document correlated parameters as "dot plots" or "contour plots", or one might prefer uncorrelated results as "histograms", which relate to a single parameter each. Dot plots are usually taken as polaroid photographs directly from the oscilloscope screen, whereby each dot on the screen corresponds to a single cell which is characterized by the coordinates of parameter 1, respectively, 2. These oscilloscope pictures are generated by the adjustable dot buffer, which displays the incoming signals on a first-in first-out basis. Thus, the dot plot will respond rapidly to variations of incoming signals, which makes it very useful for optimizing instrument adjustments or rapid overview on complex distributions. The frequency distribution of cells may also be deduced from the dot plot, since frequent events will appear as a cluster of dots. Other types of instruments use the dot-plot feature while integrating over time, producing so-called cytograms or scattergrams. However, even though this type of display may be more quantitative, its slow response to time-dependent variations makes it less useful.

Quantitative information from correlated two-parameter distributions is obtained from contour plots, which essentially consist of the two measuring parameters correlated in a plane with the numerical third parameter oriented perpendicular to that plane and its height presented by contour lines. This type of display may be read like an ordinary map, whereby the contour lines indicate the actual number of cells. Contour plots have been improved by defining contour lines as integrals, encompassing a given percentage of cells and making them independent from the actual cell number. Unfortunately, this modification requires elaborate computation and is therefore not widely used.

Uncorrelated data are presented as histograms with the respective parameter intensity as abscissa and the frequency or cell number displayed as ordinate. Again, since histograms are in most instances derived after gating on one or more parameters, this fact and the conditions employed should be indicated each time. Fluorescence (and also scatter) parameters can be displayed in linear or logarithmic scale; either fact has to be indicated. Whereas the linear display allows for the easy comparison of staining intensities, the limited scale makes it less useful for widely differing fluorescence intensities, such as those seen during normal phenotyping. In addition, weakly labeled subsets of cells are less well separated compared to the log display, which will easily keep both weakly and heavily labeled subsets as discrete peaks on scale. However, quantitation is more difficult to achieve and requires the use of internal standards, such as fluorescently labelled latex beads and the calibration of amplifiers by neutral density filters. Most important in the log display during quantitative measurements is the true Gaussian distribution of the negative (left) peak. Unfortunately, in log display some instruments only provide autofluorescence (negative) signals as spike in the first few channels, like on linear scale, thus losing resolution for weakly-stained cells and making quantitation impossible.

So far, we have only related to fluorescently labeled cells, not specifying the nature of the fluorescent label. In general, any type of fluorescent label may be introduced if it (a) matches the laser wavelength used for excitation and (b) results in a fluorescent signal which is separable from the exciting light, as well as from eventual autofluorescence. Thus, DNA and RNA may be quantitated by flow cytometry, cell-membrane fluidity may be assessed using fluorescent probes, or cell membranes can be labeled using appropriate monoclonal antibodies coupled to a fluorescent dye. Such dyes are listed in table I. Certainly most widespread is fluoresceine, which ideally matches the strong 488 nm line of an argon laser and results in bright green fluorescence. Tetramethyl-rhodamine or the rhodamine derivative, XRITC, ideally require the use of dual laser excitation to achieve optimal separation of red from green fluorescence. Recently, a sulfonylchloride derivative of rhodamine called, "texas red", has surpassed its predecessors, due to its superior spectral properties; it is used exclusively for two-laser work.

Table I. Fluorescent dyes

Fluorochrome	Excitation (nm)	Emission (nm)
Fluorescein (FITC)	489	515
Tetramethylrhodamine (TRITC)	554	573
XRITC	582	601
Texas red (TR)	595	625
Phycoerythrin B (PE)	545, 565	570
Phycoerythrin R	480, 545, 565	578
Allophycocyanine (APC)	650	660

In contrast to these low-molecular weight dyes, highly fluorescent proteins, prepared from certain algae, have recently been introduced. These phycobiliproteins are a family of proteins (molecular weight far above 100 000) which exhibit high Stokes shifts and extremely high quantum efficiencies. Phycoerythrin B is already widely used for two-color fluorescence, in combination with fluoresceine using excitation at 488 nm. Since the emission spectra of these two fluorophores partially overlap, fluorescence has to separated with appropriate dichroic mirrors. However, to get complete separation of both colors, signals have to be corrected by an electronic network similar to that used in scintillation-counting for the separation of two different isotopes. Thus, at present, two-color fluorescence for routine purposes will use fluoresceine in combination with phycoerythrin B, both excited by a single laser at 488 nm, while fluoresceine together with Texas red excited by two lasers will be the choice for sophisticated applications.

As important as the instrument set up, is the proper labeling of the cell suspension. Specificity of the label is directed by the antibody in use. However, to achieve optimal results, the background introduced by the labeling procedure should be as low as possible; the specific label should be most intense and consistent between experiments. Non-specific reactivity of the antibodies, such as interaction with Fc-receptors should be kept to a minimum. In our laboratory, we have adopted a procedure which allows a large number of samples to be processed without much effort and which keeps unwanted reactivities to a minimum. This procedure, as used for human cells, is outlined in table II. In summary, the use of microtiter plates enables rapid processing of many samples and the use of preabsorbed F(ab')$_2$-fragments of labeled secondary antibody in combi-

Table II. (Indirect) Immunofluorescence labeling procedure for standard phenotyping

Cells
Prepare human mononuclear cell suspension by standard Ficoll-Hypaque centrifugation. Suspend cells in tissue culture medium or wash buffer.

Primary Antibodies
Use commercial monoclonal antibodies undiluted or 1 : 10 in wash buffer as recommended. In case of raw culture supernatants, 10 or 20 μl are usually needed. For (direct) immunofluorescence of surface immunoglobulins, use only F(ab')$_2$-fragments. Centrifuge before use. See below.

Secondary Antibodies
The use of F(ab')$_2$-fragments is strongly recommended. Use one pretested batch, which is stored deep-frozen in suitable aliquots. This antibody preparation should be preabsorbed with human serum at a 1 : 1 ratio (v/v) for 30 min at 37° C, followed by 60 min at 4° C. Complexes are removed by ultracentrifugation for 90 min at 100 000xg and the pellet discarded. This absorption procedure may not apply when using affinity-purified antibodies. The final (working) dilution is made in wash buffer before use; thereafter, the antibody is stored at 4° C for short periods and not refrozen. Immediately before use, all primary and secondary antibodies are centrifuged for 5 min in a Eppendorf centrifuge.

Other Reagents and Materials
Wash buffer: Consists of standard phosphate buffered saline with addition of 0.1 % bovine serum albumin and 0.1 % NaN$_3$. Blocking IgG: Consists of human immunoglobulin for intravenous application such as Intraglobin (Biotest), which is diluted in phosphate-buffered saline 1 : 5 (v/v) resulting in 10 mg/ml immunoglobulin. Microtiter plates: flexible microtiter plates (e.g., Falcon # 3911) are most suitable. The use of a microplate-dispensing unit and a microplate shaker is recommended.

Procedure
– Prepare microtiter plate for the distribution of antibodies by marking the appropriate wells, preferably set up in rows of twelve, leaving alternate rows empty;
– all procedures now require to be set up on ice and all reagents should be ice-cooled;
– use 500 000 to 1 000 000 cells per well; with special precautions even less can be used. Pipette cell suspension into wells and fill with wash buffer at 4° C. Centrifuge for approx. 1 min at 1500 g. Centrifugation may be even shorter. This should be individually tested. At the end of the centrifugation, a fine pellet of cell should be visible. Supernatant liquid is removed by flicking the plate, the cell pellet is resuspended (before adding new wash buffer) on a microplate shaker and the washing procedure repeated twice;
– to the pellet of washed cells, add one drop (= 20 μl) of blocking IgG and shake again; do not add blocking IgG when assaying surface immunoglobulins;
– add appropriate primary antibody or medium control. Do not use antibody volumes greater than 100 μl to avoid spillage during shaking. Shake plate on microplate mixer;
– incubate on ice (at 4° C) for 30 min with occasional shaking;
– fill with wash buffer and wash twice by centrifugation as above. To avoid crossleakage, clean microplate surface after flicking with new dry paper towel;
– to washed cell pellet add secondary antibody dilution. Shake as above;
– incubate in the dark another 30 min on ice with occasional shaking;
– fill with wash buffer and wash at least twice as above;
– finally resuspend cells in 100 to 200 μl wash buffer, cover microplate and store dark on ice until assayed;
– direct immunofluorescence proceeds similarly with only one incubation.

nation with human IgG, added to block Fc-receptors, results in easily reproducible and specific immunofluorescence assays.

Examples and Applications

The following examples are aimed at demonstrating some of the basic principles of flow cytometric analysis. Furthermore, applications on vari-

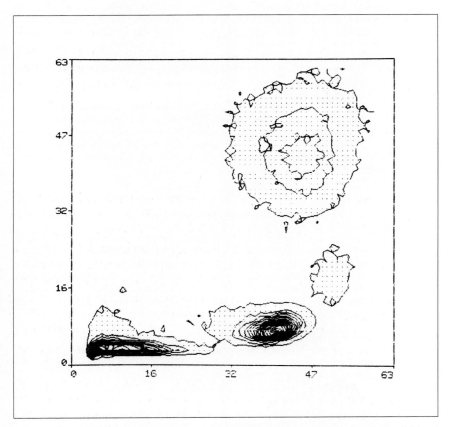

Fig. 4. Forward versus right-angle scatter distribution of human peripheral blood leukocytes: Contour plot comprising 50 000 events. Abscissa denotes forward-angle scatter (linear) and the ordinate is right-angle scatter (quasi-logarithmic). Peaks are (counterclockwise) (a) cellular debris, erythrocytes and thrombocytes; (b) lymphocytes; (c) monocytes, and (d) granulocytes.

ous clinical samples may reveal its unique capabilities:

Flow cytometry in a clinical environment usually means to process samples of peripheral blood and to some extent, bone marrow specimens. To selectively analyze the major leukocyte subpopulations, samples may be enriched for leukocytes by standard dextran sedimentation techniques and/or lysis of red blood cells. Alternatively, when only mononuclear cells are required, one may use the standard procedure of Ficoll-Hypaque centrifugation. Such samples are labeled as described above and analyzed on a flow cytometer. With appropriate instrument settings, such peripheral blood leukocytes are analyzed for their forward versus right-angle scatter distribution, as shown in the contour plot in figure 4. The abscissa indicates in linear scale the intensity of forward angle light scatter, i.e., cell size and the ordinate depicts right angle scatter which informs us about the internal structure of the cells. The special instrument settings reveal a non-linear scale in the ordinate, which is required to keep all cell types on scale. As can be easily deduced, the contour plot defines four clusters of cells (listed counterclockwise):

– Debris, erythrocytes, and platelets
– lymphocytes
– monocytes
– granulocytes.

Thus, the two scatter parameters without any additional fluorescence labeling allow complete separation of the major leukocyte subsets found in peripheral blood and also (not shown) in bone marrow. Additional peaks may show up in cases of hematological malignancies. From our technical considerations above, it may be readily inferred that gating on one of these clusters will result in selective analysis of the fluorescent parameters of, e.g., lymphocytes despite the presence of a vast majority of other cell types. This is particularly helpful when phenotyping lymphocyte subsets from a small volume of whole blood. Furthermore, it is clear from figure 4, that gating on forward scatter alone will not result in "pure" cell populations (e.g., lymphocytes free of contaminating monocytes). Forward versus right-angle scatter distribution (dual scatter; sometimes referred to as scattergram) is particularly important when analyzing leukemia patients' samples: The monoclonal proliferation of leukemia cells may manifest itself as a discrete cluster of cells, again only defined by dual scatter. If so, the use of dual-scatter gating allows the selective analysis of the leukemic cell phenotype in parallel to the patients' normal lymphocyte subsets. When only limited material is available, recording of data in list-mode allows repeated evaluation of the same set of data for optimal results.

The example presented in figure 5 illustrates how gating can be used to visualize even minor cell subsets, provided a characteristic marker for this subset is available. We call this technique fluorescence-associated scatter (FAS) gating. Thus, to localize by dual scatter a cell subset which carries a specific marker, gating is performed on the appropriate fluorescence level and the resulting scatter distribution recorded as a dot plot. Figure 5 (left) shows the forward versus right angle scatter distribution of normal human peripheral blood mononuclear cells. The cluster of lymphocytes is clearly visualized in the middle of the dot plot; its electronic gate is indicated by the four markers (amplifier settings are different from those in fig. 4). There are few monocytes, as seen from the dots to the right and above the lymphocytes. In order to visualize the non-specific uptake of immunoglobulin via Fc-receptors, this sample was prepared without blocking immunoglobulin. After establishing the ungated dual-scatter distribution, the sample was rerun on the FACS, however, with a gate set on the fluorescence parameter, so that only cells exceeding a given threshold fluorescence would be analyzed for their scatter parameters. The resulting fluorescence-associated scatter distribution is shown in figure 5 (right). As can be seen, monocytes have selectively taken up immunoglobulin by their Fc-receptors under such conditions, whereas the

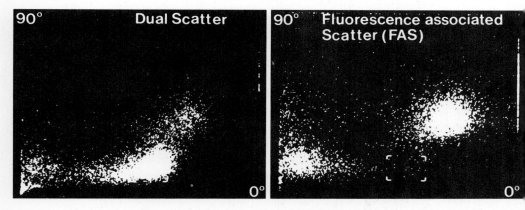

Fig. 5. Dual-scatter distribution of human peripheral blood mononuclear cells (left) and principle of fluorescence-associated scatter gating (right). Dot plot where each point represents one cell; total number of cells displayed is 4032.

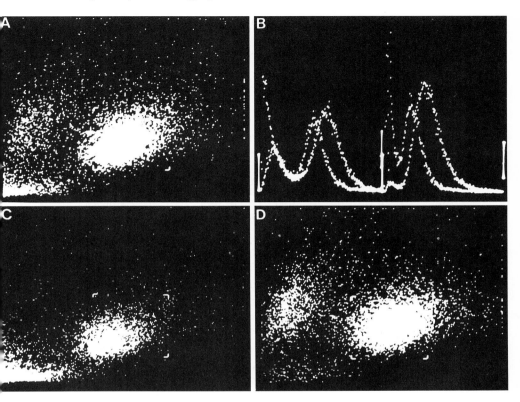

Fig. 6. Subpopulations within one "homogeneous" cell line. Cells are cultured human lymphoblasts, transformed in vitro by Epstein-Barr virus. (A) Dot plot of forward vs. right-angle scatter distribution (compare fig. 4) of cell line labelled with OKT9 (transferrin-receptor). (B) Normalized histograms of FAS-distributions from figures 6c and 6d. Left panels are forward scatter and right panels are right-angle scatter. (C) Fluorescence-associated scatter (FAS) distribution of OKT9-negative cells. (D) FAS pattern of OKT9-positive cells.

lymphocyte scatter gate (visualized by the four markers) is virtually free of cells showing positive fluorescence. This approach is universally applicable and can be used to localize (in terms of scatter) subsets of normal and pathological cells. It may also be used to more precisely define the dual-scatter gates to be used for scatter gating.

To demonstrate the resolving power of this approach, the forward versus right-angle light-scatter distribution of a human B-lymphoblastoid cell line (transformed in vitro by Epstein-Barr virus) is shown in figure 6a.

The cluster of lymphoblasts is set to the center of the display by adjusting amplifier gains and appears very homogeneous. This cell line has been labelled with various monoclonal antibodies, one of them directed against the receptor for transferrin (OKT 9). The fluorescence profile for this marker is clearly bimodal, which may be explained by the fact that proliferating cells express transferrin receptors to cover their supply in iron. We may next ask whether these presumed proliferating cells differ in any respect from their non-proliferating counterparts? Thus, fluorescence-associated scatter-gating is applied to the transferrin-receptor-negative (fig. 6c) and to the transferrin-receptor-positive (fig. 6d) cells. The cells which do not carry transferrin receptors form a narrow cluster within the total distribution and are distinctly smaller than the remainder of the population. In contrast, transferrin-receptor-positive lymphoblasts are more heterogeneous in size and larger than their resting counterparts. Figure 6b demonstrates this fact by overlapping the corresponding histograms from the same experiment: The two histograms on the left clearly show the difference in forward scatter, whereas the histograms on the right relate to right-angle scatter. Thus, within this apparently homogeneous cell line, differences in surface markers can be related to characteristic variations in cell size and structure using fluorescence-associated scatter. This example, from an in vitro cultured cell line, is paralleled by many specimens from patients with apparently "homogeneous" leukemic cell populations. Here, not only the state of proliferation may vary as assessed by the presence or absence of transferrin receptors, but also other surface markers may show considerable alterations between resting and blast cells. Testing for such changes will help our understanding of leukemic-cell differentiation and improve monitoring of patients by phenotypic analysis.

In most instances, interest is in the analysis of peripheral blood lymphocytes to assess phenotypically the state of the immune system. This is done using the appropriate antibodies in indirect immunofluorescence, as described above, and analyzing the lymphocyte cluster after gating on forward versus right-angle scatter. The resulting histograms are shown in figure 7. They are positioned above each other to allow comparison of fluorescence intensities and distribution patterns. The fluorescence scale in the abscissa covers 2.5 decades in logarithmic scale. Cell number is shown in the ordinate and each experiment is performed on 20 000 lymphocytes. Needless to say, counting fluorescently labelled cells under a microscope will be less accurate. The histogram pattern and the respective

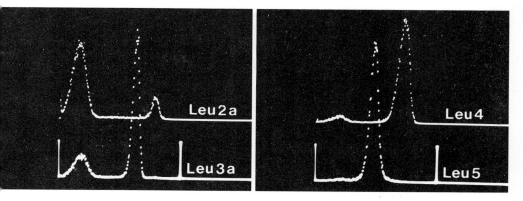

Fig. 7. Histogram pattern of "standard" T-lymphocyte markers in human peripheral blood. Horizontal fluorescence axis extends between vertical markers. Scale is logarithmic in 2.5 decades. Each histogram represents 20 000 cells. The peak on the left represents unstained cells (autofluorescence).

fluorescence intensities are typical for each of the antibodies shown: Antibody Leu2a against T-lymphocytes of suppressor/cytotoxic phenotype is easily recognized by a small intensively labelled subset, shown in the peak on the right, and a minor, but varying, subpopulation of intermediately labelled cells extending between the negative and the strongly positive cells. This relationship is documented also by two-color fluorescence in figure 12a. Antibody Leu3a defines helper/inducer cells. Its histogram is recognized even in coded samples by its narrow fluorescence distribution. Leu4 recognizes an epitope present on all mature T-lymphocytes, while antibody Leu5 is directed against the receptor for sheep erythrocytes and thus represents the broadest marker for human T-lymphocytes. As already pointed out, each of these monoclonal antibodies will define a particular subset of cells and – as long as standard conditions are employed – will stain those cells with the same intensity. This is most evident when analysis is done with fluorescence in log scale. Here the antibodies described above, and most others used for phenotyping, have their own characteristic histogram pattern, which is easily distinguishable from the others. This even provides an internal control on the proper setup of the labeling experiment. Alterations from these patterns indicate abnormal antigen expression in terms of antigenic determinants per cell, although the proportion of cells may still be within the normal range. Such quantitative changes in antigen expression may occur on malignant cells or may reflect other, functional parameters of the population under study. At present, we are only at the beginning of our understanding of such subtle variations. However, from what is known today, the quantitative evaluation of immunofluorescence patterns will provide important insights into the functional state of the immune system.

One word of caution has to be added. Since most of the monoclonal antibodies, which are currently available, define surface structures on cell subsets which have to be independently defined as to their function, we should always restrain our issue to the "phenotype" of that cell. For example, unless the suppressive activity of a cell population has not been demonstrated, we should talk only on "suppressor-phenotype cells", since phenotype and actual function might be quite divergent.

Phenotyping of homogeneous samples is easily performed from dot plots as shown in figure 8. Here, peripheral blood mononuclear cells from a patient with chronic lymphocytic leukemia (CLL), were analyzed for surface immunoglobulin and other differentiation markers. Displayed are dot plots with forward scatter in the abscissa and log fluorescence (2.5 decades) in the ordinate. The gate markers relating to forward scatter indicate the position of the lymphocyte cluster. As can be seen, the patients' cells possess surface IgG with a monoclonal light chain (kappa). However, shifts in the peak for lambda light chain indicate that also some serum immunoglobulin is adsorbed via Fc-receptors. In contrast, IgD is completely negative on those leukemic cells. When the monoclonal antibody OKM1 (directed against monocytes, granulocytes, null-cells) is applied to the same sample, two clusters of cells are resolved. The majority of cells is positive for that marker, as opposed to a small cluster of dots which do not bear the OKM1 marker. Since the analysis of surface immunoglobulin has revealed that most lymphocytes in that sample belong to the neoplastic population, the leukemia cells of that patient may be considered OKM1-positive. The nature of the OKM1-negative cells is readily apparent when the marker OKT3 (against all mature T-lymphocytes; analogous to Leu4) is applied: Here a very small cluster of dots is positive, indicating that these few cells are the residual normal T-lymphocytes in that patient. Similar conclusions can be drawn when two anti-

Fig. 8. Phenotyping of human chronic lymphocytic leukemia cells: Dot plots represent forward scatter versus log fluorescence (2.5 decades). Antibodies used are specified in the upper right corner.

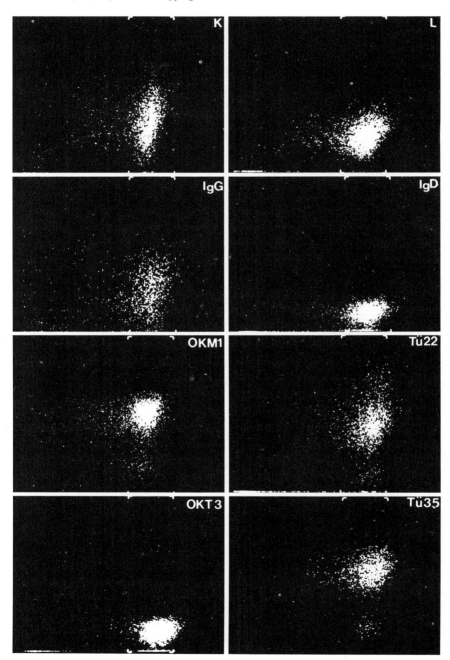

bodies against HLA-DR- and -DC-antigens are used: The monoclonal antibody TÜ22 (against HLA-DC), results in three clusters of cells: (i) A negative population corresponding in number to the normal T-lymphocytes, (ii) the majority of cells being labeled at intermediate intensity, and (iii) a small subset of cells with even higher antigen density, which indicates that according to that marker, the leukemia cells may be considered heterogeneous. In contrast, the monoclonal antibody TÜ35 (against HLA-DR) uniformly reacts with the leukemic cells and is non-reactive with the normal T-lymphocytes. This differential reactivity of several markers may help to position a leukemic phenotype within the B-cell differentiation pathway.

It should be noted, however, that such single fluorescence analysis will not unequivocally establish these presumed correlations. The interpretation and correlation of markers proposed here for this phenotyping experiment is based on plausibility – improvements on this will only be feasible when two-color fluorescence assays are performed, as will be shown later.

The following examples demonstrate that changes in quantitative antigen expression indeed occur and are worth noting. We have performed phenotypic analysis on a number of human cord blood samples. A standard result is shown in figure 9a with two histograms (in log; 2.5 decades), derived from monoclonal antibodies Leu4 (small dots) and Leu3a (large dots) superimposed. Those histograms were derived by dual-scatter gating on 20 000 lmyphocytes. The relative intensities of both markers are within the normal range. Figure 9b shows one sample from a pair of monozygotic twins with the same markers. The profile on the right representing the pan-T-marker Leu4 is essentially unchanged. However, the quantitative expression of the Leu3a determinant is strongly reduced, as on the cells of his newborn brother (data not shown). In contrast, the percentage of labelled cells is within normal range. This unique finding of reduced quantitative expression of the Leu3a antigen remains, so far, unexplained. However, other reports have shown that during mixed leukocyte culture in vitro, the quantitative expression of the Leu3a-analogous structure T4 decreases. This phenomenon has been ascribed to the influence of interferon-alpha. If applicable to our newborn twins, it might point to the fact that some of the twins' T-cells were activated already in utero. Our current attempts are at defining such influences and their potential prognostic significance.

More obvious and rather self-explanatory are alterations in quantita-

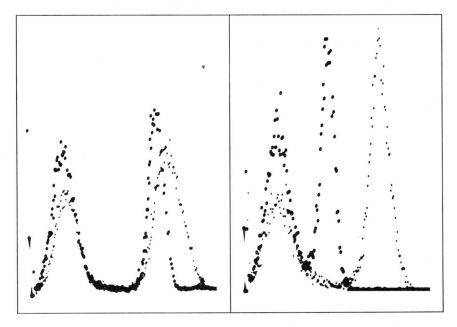

Fig. 9. Quantitative variations in antigen expression: Leu3a versus Leu4. *a* Histograms of normal human cord blood lymphocytes stained with anti-Leu4 (small dots) and anti-Leu3a (large dots); *b* same with cord blood lymphocytes of newborn twin.

tive antigen expression induced by therapeutic manipulations. Figure 10 depicts an example of a boy (JH) with endogeneous eczema and multiple warts, who would not respond to conventional therapy. He was treated with an infusion of interferon-alpha over 4 days. Since we had identified an abnormal lymphocyte population in his peripheral blood (see below fig. 12b), his peripheral blood mononuclear cells were monitored by phenotypic analysis in short intervals. To demonstrate the changes in quantitative antigen expression which might occur during immunomodulatory treatment, the results of labeling for HLA-DR (antibody 7.2) are presented. Similar results were obtained using a variety of other MHC-specific antibodies. Pretreatment fluorescence distribution for this antibody (fluorescence in log scale, 2.5 decades, dual-scatter gating on 20 000 lymphocytes) is shown in the top panel. This is essentially indistinguishable from a normal control. The middle panel recorded 6 days after beginning of the treatment shows that HLA-DR is still present on this patient's B-cells, however at strikingly reduced quantitative levels. Looking at the same

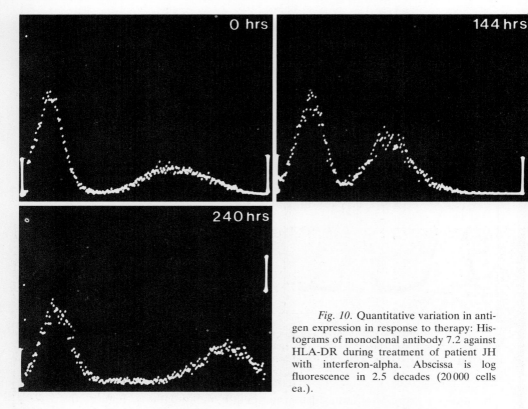

Fig. 10. Quantitative variation in antigen expression in response to therapy: Histograms of monoclonal antibody 7.2 against HLA-DR during treatment of patient JH with interferon-alpha. Abscissa is log fluorescence in 2.5 decades (20000 cells ea.).

patient after 10 days, we realize that quantitative antigen expression has changed again. This time, however, HLA-DR-levels are increased and supercede even the pretreatment values. Most important is the fact that the percentage of those cells has not changed appreciably during that time. Further follow-up indicated later a reversion to the normal pretreatment values. These results are at variance with observations on interferon-treatment in vitro: Even though it is known that HLA-A,B,C-antigens are increased in response to interferon-alpha, HLA-DR is not. In contrast HLA-DR is thought to be regulated by interferon-gamma. One might speculate that these changes, which are evidently in response to treatment, reflect other, secondary alterations in the interferon system, e.g., by raising endogeneous interferon-gamma levels which then lead to the observed changes. Notwithstanding, these findings underscore the value of quantitative immunofluorescence assays in phenotypic analysis, especially during experimental immunotherapy.

Some instances are already known where quantitative antigen expression is related to differentiative events. Figure 11 shows in comparison the surface-IgM- and -IgD-profiles of human cord blood lymphocytes and adult lymphocytes. All histograms are in log scale and have been acquired under comparable conditions. Even though the number of B-lymphocytes in adults is usually much lower compared to cord blood, the differing histogram patterns of surface-IgM between both sources are evident. A rather small difference is noted in the density of IgD. It is known from oncogenetic studies in the mouse, that B-lymphocytes will first express

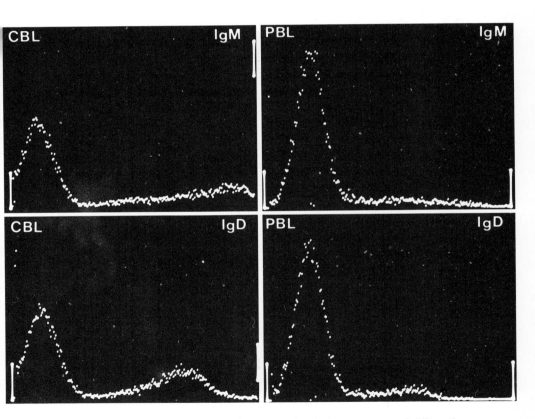

Fig. 11. Quantitative variation in antigen expression during ontogeny and differentiation: Surface IgM (top panel) and surface IgD (bottom panel) on cord blood (CBL) and adult lymphocytes (PBL). Abscissa is log fluorescence in 2.5 decades.

IgM alone, then co-express IgD while IgM-expression reaches a maximum and then declines. Thus, by careful analysis of the quantitative expression of surface immunoglobulins, one might derive the exact position of a cell within the B-cell differentiation pathway. Similar findings apply to the situation in man. However, as opposed to the experimental system in the mouse, there are no other markers available to characterize those differentiative stages. This may be of clinical relevance, since those cord-blood-type B-lymphocytes are apparently lost from the circulation within the first year of life. We have observed, however, a group of immunodeficient patients, completely lacking the expression of MHC-class-II-antigens, which retain this immature pattern beyond the age of one year. This clearly indicates that no further B-cell maturation has taken place in these rare patients. It further implies that for normal B-cell differentiation to occur (at least for differentiation beyond a certain stage), the presence of HLA-D-region antigens on the interacting cells of the immune system is mandatory. This important information was only derived by analyzing the fluorescence-intensity distribution on the patients' B-lymphocytes.

As already mentioned, quantitative immunofluorescence using a single fluorescence label may not be sufficient to completely define the complex phenotype of a given cell. In particular, if it is not distinguishable from other cell types by means of forward-versus-right-angle scatter, or if it displays a phenotype which is not usually found on peripheral blood or bone marrow cells. In such cases, definition of a complex surface-marker phenotype may be obtained with a two-color fluorescence system. The results of such investigations are usually displayed as dot plots or contour plot to visualize the correlated expression of the two surface markers. Figure 12a gives an example of normal peripheral blood lymphocytes selected by dual-scatter gating and analyzed for two different T-cell markers. The abscissa denotes log fluorescence obtained by directly fluoresceinated monoclonal antibody Leu1, which labels most T-cells, a small subset of B-cells and leukemic cells from patients with chronic lymphocytic leukemia. The ordinate denotes suppressor phenotype T-cells, as defined by monoclonal antibody Leu2a, which was used as biotinylated derivative and developed with Texas red conjugated to avidin in the second incubation. All scales shown are in log (2.5 decades, 64 channels) and cells defined negative are found between channels 0 and 16. Experiments were done on 50 000 lymphocytes. The lymphocytes of a healthy individual (fig. 12a) are clearly separated into three major groups: As expected, most cells bear the marker Leu1, i.e., most cells in normal

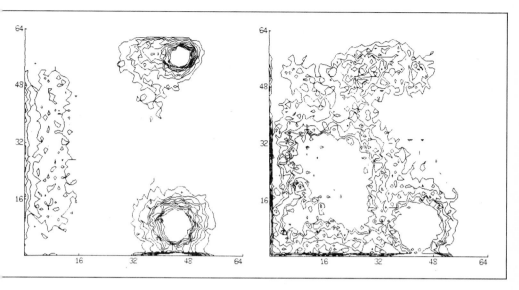

Fig. 12. Two-color fluorescence. Contour plots of two-color fluorescence distributions (gated on forward vs. right-angle light scatter). Displayed are 50 000 cells. Abscissa denotes FITC-labeled (Leu1 (pan-T cell marker). Ordinate is biotinylated Leu2a antibody against suppressor-phenotype cells, developed with Texas red-avidin. Figure 9a is the normal control and figure 9b is patient JH.

peripheral blood are T-lymphocytes. Among T-cells, the clear distinction between the heavily labeled suppressor-phenotype cells and the cells negative for that marker (presumably the remaining helper/inducer cells), is most evident. Interestingly, all cells bearing the Leu2a marker at intermediate density are in the Leu1-negative fraction. This relates to the histograms in figure 7, where we noted that the Leu2a intensity may identify different subsets. However, with two-color fluorescence this distinction is readily apparent. The same setup of markers is applied to peripheral blood lymphocytes from the patient (JH) introduced in figure 10 and depicted in figure 12b. This highly complicated contour plot allows a number of conclusions: Apart from the cell populations found in a normal individual (see fig. 12a), there is a major intermediate population which carries both the Leu1 and the Leu2a markers. This population was found in other experiments also carrying surface immunoglobulin and HLA-D-region antigens, thus relating to a cellular phenotype like that found in chronic lymphocytic leukemia. However, in this young patient there was no indication of a lymphoid neoplasm. Moreover, comparing the fluorescence intensities indicates that the Leu2a strongly positive population carries less Leu1-antigen compared to the normal control. This may indicate that the suppressor phenotype cells in that patient are different from those normally found.

Despite the enormous amount of information which can be drawn from such studies, this type of analysis is still rather complicated and should be reserved for special applications. However, in such instances, it will provide information superior to any other method. In addition, the direct demonstration of such unusual combinations of surface markers will help us to better understand the complexity of the immune system.

Conclusions

To summarize, we have shown that the phenotypic analysis of leukocyte subsets using flow cytometry provides a powerful means of differentiating single cells in suspension. In combination with the resolving power of the analysis of scattered light, it provides a well-defined, reproducible approach for classifying cell populations. The computing facilities of a flow-cytometric instrument will enable exact evaluation of any cell type currently defined by the appropriate reagents. The speed and accuracy of that approach is, by order of magnitude, superior to the counting of cells on a fluorescent microscope. In contrast to the rather subjective microscopic evaluation, flow cytometry renders immunofluorescence amenable to quantitative measurements. Thus, apart from the enumeration of cell types, quantitation of antigen density on single cells is now feasible. At present, we are still ignorant of most of the variations observed. However, the wealth of information provided will soon contribute to a better understanding of the physiology of the immune system in health and disease.

Acknowledgment

The cooperation of Drs. R. Dopfer, A. Fridrich, C. Müller and J. Treuner is greatfully acknowledged. It is a pleasure to acknowledge the technical assistance of Ms. M. Wiedmann. Monoclonal antibodies were from commercial sources (Becton-Dickinson, Ortho, New England Nuclear) or were the kind gift of Dr. A. Ziegler, Tübingen, FRG (TÜ 22, TÜ 35). All flow cytometric analyses were performed on a FACS-IV cell sorter at the FACS-laboratory of the Medizinische Klinik in Tübingen, FRG. Contour plots were generated on a Vax-11 computer at the Max-Planck-Institut für Biochemie, Martinsried, FRG, with invaluable help and programs provided by Dr. L. Voet, Martinsried. Contour plots in figure 12 were generously prepared by Dr. F. Nauwelaers, BD-FACS-Systems, Sunnyvale, Ca., USA, on a Consort-40 computer system. This study was supported in part by DFG (SFB 120/F1) and the Robert-Bosch-Foundation.

Summary

The paper describes the basic setup of a flow-cytometric instrument with focus on the various applications of multi-parameter analysis. It will introduce the novice reader to some technical details relevant to the understanding and proper interpretation of flow-cytometric data presentations. To complement the technical aspects, a detailed procedure for the immunofluorescent labeling of cell suspensions is described. The applications discussed for the phenotypic analysis of human peripheral blood leukocyte subsets include forward-versus-right-angle scatter analysis, fluorescent-associated-scatter gating, and different examples for the use of quantitative immunofluorescent measurements. Finally, the superior resolving power of two-color immunofluorescence is demonstrated.

Zusammenfassung

Die Arbeit beschreibt den grundlegenden Aufbau eines Durchflußzytofluorimeters mit einem besonderen Gewicht auf den verschiedenen Anwendungen der Multiparameter-Analyse. Sie soll den Nichtspezialisten mit einigen technischen Details vertraut machen, die für das Verständnis und die richtige Interpretation von durchflußzytofluorimetrischen Daten unerläßlich sind. In Ergänzung dazu wird eine einfache und schnelle Methode zur Fluoreszenz-Markierung von Zellsuspensionen beschrieben. Die Anwendungen bei der Phänotypisierung von humanen Leukozyten-Subpopulationen umfassen die Analyse von 0°/90°-Lichtstreuung, Fluorescence-associated-scatter Gating, und verschiedene Beispiele für den Wert quantitativer Immunfluoreszenz-Messungen. Den Schluß bildet eine Demonstration der hohen Auflösung phänotypischer Untersuchungen mit Hilfe der Doppelimmunfluoreszenz-Methode.

References

In place of an extensive bibliography, the reader is referred to a few reviews on the general subject of Flow Cytometry and its applications. References to other topics covered in this article may be obtained from the author.

Melamed, M. R.; Mullaney, P. F.; Mendelsohn, M. L. (eds.): Flow Cytometry and Cell Sorting (J. Wiley, New York 1979).
Greaves, M.; Delia, D.; Newman, R.; Vodinelich, L.: Analysis of Leukaemic Cells with Monoclonal Antibodies; in McMichael, Fabre (eds.), Monoclonal Antibodies in Clinical Medicine, pp. 129–165 (Academic Press, New York 1982).

Beverly, P. C. L.: The Use of the Fluorescence-Activated Cell Sorter for the Identification and Analysis of Function of Cell Subpopulations; in McMichael, Fabre (eds.), Monoclonal Antibodies in Clinical Medicine, pp. 557–584 (Academic Press, New York 1982).

Kruth, H. S.: Flow Cytometry: Rapid Biochemical Analysis of Single Cells. Anal. Biochem. *125:* 225–242 (1982).

Dr. Martin R. Hadam, Onkologisches Labor, Abt. Kinderchirurgie, Kinderklinik, Med. Hochschule Hannover, Konstanty-Gutschow-Straße, D-3000 Hannover 61 (FRG)

Clinical Studies

Xenogenic Peptides and Proteins in Myelo- and Lymphoproliferative Disorders

H. Röhrer

Internal Dept., Rosmann Hospital, Breisach, FRG

The bone-marrow and the lymphatic system produce various cell systems with varying specific functions. Recent studies have shown that myeloid and lymphatic cells have a common stem cell, known as a pluripotent cell, which is located in the bone-marrow but can also be recruited from other tissues. From pluripotent stem cells in bone-marrow, there develop the so-called determined stem cells for the erythropoiesis, thrombopoiesis, monocytopoiesis and granulopoiesis. In the thymus, the immigrant pluripotent stem cells develop into determined stem cells of the T-lymphocytes and in the lymphatic bursa equivalent to determined stem cells of the B-lymphocytes. There is a series of development stages between the determined stem cell and the mature definitive cell appearing in the blood. This may be illustrated again by the example of granulopoiesis: in this developmental process, we differentiate, in this order, between myeloblasts, promyelocytes, myelocytes, metamyelocytes, stab cell, and polymorphic granulocytes. The myeloblasts and some of the promyelocytes represent the proliferation pool; promyelocytes, myelocytes and metamyelocytes the maturation pool; stab cell and polymorphic granulocytes the storage and function pool. In principle, there is a positive feedback mechanism between the pluripotent stem cell and the function pool, this involves, in addition to various factors which are still unknown, lymphokines and monokines such as the colony-stimulating factor (CSF) and the granulocyte/macrophage differentiation factor (GMF).

A biological system of such complexity confronts pathologists, immunologists, and clinicians with benign and malignant disorders of similar complexity. The following illustrations (table I–IV) provide an overview of the classification of myelo- and lymphoproliferative disorders.

In this connection, important new prognostic and therapeutic aspects have been supplied by the FAB classification of acute myeloid leukemias (classification of the French-American-British Co-Operative Group) which is elaborated on a morphological and cytochemical basis. This is especially valid for immunological classification of acute lymphatic leukemias by means of monoclonal antibodies.

According to the immunological classification of acute lymphatic leukemias, the common ALL (c-ALL) and the null-ALL represent an accumulation of lymphoid precursor cells, whereas, with the T-ALL and the B-ALL, it is partially the matured T- or B-lymphocytes which accumulate. The c-ALL of the child, representing the most common form of leukemia in this age, displays a particularly good response to the therapy and, as is well-known, has a favorable prognosis [1].

Very recent immunological and proliferative kinetic studies of myeloid and lymphatic leukemias have clearly demonstrated that a leukemia cell is not a "monster cell" with uncontrolled proliferation, but a cell which, for reasons still unexplained, is more or less blocked in its maturation to a function cell. The blockade begins on the plane of the

Table I. Types of myeloproliferative disorders

(1) Acute	Di Guglielmo syndrome
	Myelofibrosis with myeloid metaplasia
	Erythremic myelosis
	Erythroleukemia
	Acute myeloid leukemia
(2) Chronic	Polycythemia vera
	Myelofibrosis with myeloid metaplasia
	Chronic myeloid leukemia
	Essential thrombocythemia

Table II. FAB Classification of acute myeloid leukemias

M1	Myeloblastosis without maturation
M2	Myeloblastosis with signs of maturation
M3	Promyelocytic leukemia
M4	Myelomonocytic leukemia
M5	Monocytic leukemia
M6	Erythroleukemia

Table III. Types of lymphoproliferative disorders

(1) Acute	Acute lymphatic leukemia (c-ALL, T-ALL, B-ALL, Zero-ALL)
(2) Chronic	Chronic lymphatic leukemia (B-cell-T-cell-T/B-cell leukemia)
	Sezary's syndrome (T-helper-cell leukemia)

Table IV. Immunological classification of ALL

	c-ALL	T-ALL	B-ALL	Null-ALL
ALL-antigen	+	±	–	–
Ia-antigen	+	–	+	±
T-cell antigen	–	+	–	–
E-rosettes	–	±	–	–
Superficial immunoglobulins	–	–	+	–
TdT	+	+	–	+

stem cell. Through the blocking of the maturation process, there is an enormous prolongation of the life of such a cell, which leads to the accumulation of these cell elements in the bone-marrow, lymphatic system and blood. This accumulation subsequently leads to a suppression of normal stem cells, mechanical displacement not being the only cause. The extent of the maturation-block is more marked in acute myelo- and lymphoproliferative disorders, which against the background of these new findings, should be more aptly named myelo- or lympho-accumulative disorders, than in the chronic forms.

The cell-doubling times quoted in table V clearly indicate this direction.

These findings have led very recently to numerous in-vitro and in-vivo experiments, aimed at bringing leukemia cells to maturation. cAMP inducers such as prostaglandin E and cholera toxin, vitamin A*- and vitamin D**-analogues and lipopolysaccharides of gram-negative bacteria have displayed certain reproducible effects, especially in cell cultures. Lipopolysaccharides display an indirect effect, as they induce the formation of various lymphokines and monokines, especially CSF, after injection into animals or man. When the serum of animals or persons treated in this manner is added to leukemia-cell cultures, a clear maturation tendency of such cells occurs [2, 3].

* 13-cis-retin acid
** 1,25-dihydroxyvitamin D_3

Table V. Cell-doubling times in hours

Normal granulopoiesis	54.2 h
Chronic myelosis	64.1 h
Acute myelosis	149.0 h
Plasmacytoma	540.0 h

The natural hematological effects of injected lipopolysaccharides have long been known. According to latest findings, the time-shifted increase in myeloid and lymphatic cell elements in the blood is caused, on the one hand, by a release from the storage pool and on the other, by the induction of lymphokines and monokines with an hematopoietic effect.

An overview of the presently-known lymphokines and monokines, their source- and target-cells and functions is given in table VI.

For some time now, we have been testing xenogenic peptides and proteins, for the time being still in the form of unpurified mixtures, in respect of their efficacy as proliferation and differentiation-inducers for lympho- and myeloproliferative disorders in clinical practice. The following facts constitute the theoretical background for this:

(1) Xenogenic peptides and proteins are potent paramunity inducers

Table VI. Overview of presently-known lymphokines

Lymphokine	Source cell	Target cell	Function
Interleukin 1	Monocytes/ Macrophages	T-cells	Activation
Interleukin 2	T-cells	Cytotoxic T-cells	Growth of T-cells
Interleukin 3	T-cells	Cytotoxic T-cells	Growth of T-cells
Interferon-gamma	T-cells/null cells	Numerous	Numerous
Colony-stimulating factor (CSF)	T-cells/ Macrophages	Granulocytes	Stimulation of granulo-cytopoiesis
B-cell growth factor	T-helper cells	B-cells	Growth stimulation
T-cell substitution factor	T-helper cells	B-cells	Immunoglobulin production
Tumor necrosis factor (TNF)	Macrophages	Tumor cells	Disintegration
Lymphotoxin	T-cells	Tumor cells	Disintegration

and consequently, also stimulate the formation of lymphokines and monokines.

(2) In cell cultures, they exercise a direct effect, stimulating proliferation and differentiation, on cells exhausted by division.

(3) In cell cultures, they inhibit malignant degenerate cells but stimulate normal cells.

The possible sites of action in lympho- and myeloproliferative disorders are shown in table VII.

Since we only use these substances to treat patients who cannot or do not wish to be treated by conventional methods, the number of cases is still relatively small. The period of treatment is also still relatively short in some cases, so that at the present time, I can only report initial trends. Nevertheless, these appear to be very interesting and promising, especially since they include two cases who failed to respond to conventional therapy.

Two patients with chronic lymphatic leukemia have been under observation for 6 months; one with a primary chronic osteomyelofibrosis for 14 months and one patient with a lymphoid blastic crisis, displaying multiple chemoresistance, of a chronic myeloid leukemia. The observations made so far are as follows:

With the two patients suffering from chronic lymphatic leukemia (CLL), it is a question in one case of a CLL of the T-cell type, and in the other of a CLL of the B-cell type. For the last 6 months, both have been treated with xenogeneic peptides and proteins from animal foetal and juvenile thymus. In both cases, we have observed a normalization of the anemia which had already been present before the therapy, a regression in the swellings of the lymph-nodes and a rise in the gammaglobulin from an initial 8% and 10% to the present 12% and 16%, respectively. At the same time, there is a clearly improved mitogen stimulation (PHA and PWM) of the lymphocytes. A trend to normalization is also apparent in the lymphocyte sub-populations. With the CLL of the T-cell type, there

Table VII. Possible sites of action of xenogenic peptides with lympho- and myeloproliferative disorders

Induction of regulative lymphokines and monokines
Direct proliferation and differentiation impulses on leukemic stem cells
Inhibition of leukemic stem cells with simultaneous stimulation of suppressed normal stem cells

were initially 90% T-lymphocytes, 10% B-lymphocytes and a T-helper cell/T-suppressor cell quotient of 0.7 and after months of therapy 82% T-lymphocytes, 18% B-lymphocytes, and a T-helper cell/T-suppressor cell quotient of 1.2.

With the CLL of the B-cell type, there were initially 60% T-lymphocytes, 40% B-lymphocytes and a T-helper cell/T-suppressor cell quotient of 1.0; after 6 months of therapy 70% T-lymphocytes, 30% B-lymphocytes, and a T-helper cell/T-suppressor cell quotient of 1.3 (normal values: 80% T-lymphocytes, 20% B-lymphocytes, T-helper cell/T-suppressor cell quotient of 1.43). Up till now, there has not been a significant change in relation to the pretreatment value in the total leukocyte count of either of the two cases.

A third patient with an histologically confirmed primary chronic osteomyelofibrosis, who had had a splenectomy followed by an 1-year treatment with busulfan, has been treated for 14 months with xenogenic peptides and proteins from foetal and juvenile bone-marrow of animal origin. At the beginning of the treatment, a high-grade anemia (Hb 5 g/%, ery 1.7 million), a leukopenia of 2,000 leukocytes/ml and a thrombopenia of 8,000 thrombocytes/ml were observed. Conventional therapy was not possible in this case. Under the treatment described above, a progressive normalization of the hematological parameters took place. For 3 months, Hb 15 g/%, Er 4.5 million, leukocytes 9,000 with regular differential distribution, thrombocytes 200,000. A histological control examination no longer revealed any sign of an osteomyelofibrosis.

A fourth patient with a lymphoid blastic crisis, displaying multiple chemoresistance, of a chronic myeloid leukemia, has been under treatment for 4 months with xenogenic peptides and proteins from fetal and juvenile liver of animal origin. At the start of the treatment, the following differential blood picture was found: 62% blast cells of the T-cell type, 4% myelocytes, 4% metamyelocytes, 4% polymorphs, 10% basophils, 6% lymphocytes, 4% normoblasts; after 3 months, 13% blast cells, 1% promyelocytes, 24% myelocytes, 4% metamyelocytes, 7% stabs, 11% polymorphs, 18% basophils, 22% lymphocytes and 4% normoblasts.

At the same time, a rise in the Hb to 9 g/% and in the Er to 3.7 million (initially Hb 7 g/%, Er 2.9 million) is to be noted.

In my opinion, these observations are a pointer to certain proliferation- and differentiation-stimulating effects of xenogenic peptides and proteins on lympho- and myeloproliferative syndromes, so that a continuation with this concept in experimentation and in clinical practice appears to be justified. In the future, priority attention should be paid to the following questions:

(1) Which are the effective peptides or proteins within a mixture which, up till now, have been heterogeneous?

(2) What effects can be achieved in vitro and in vivo with purified fractions of this kind?

(3) Is there a synergistic effect with other differentiation-inducers such as 13-cis-retin acid or 1.25-dihydroxyvitamin D_3, for example?

Summary

There are experimental studies in myelo- and lymphoproliferative diseases which, contrary to solid tumors, serve as an example that through numerous heterogeneous substances (DMSO, vitamin D, vitamin A- analogues, etc.), malignant cells can be stimulated to differentiate, in some cases even be brought to a total differentiation state. There are preliminary results, which indicate a differentiation to the differential behavior-stimulus on malignant cells under therapy with xenogenic peptides and proteins. The well-known, extremely small toxicity would suggest, that patients with diseases who were unsuccessfully pre-treated with all the usual chemotherapies, should be the initiators of therapeutical studies using these substances. Documentation of four clinical cases under therapy with xenogenic peptides and proteins shows that the attempt for more research activity in this direction is justified.

Zusammenfassung

Im Gegensatz zu soliden Tumoren gibt es bei myelo- und lymphoproliferativen Erkrankungen experimentelle Untersuchungen, die belegen, daß durch zahlreiche heterogene Substanzen, wie z. B. DMSO, Vitamin D-/Vitamin-A-Analoga usw., die malignen Zellen zur Differenzierung, in manchen Fällen sogar zur Ausdifferenzierung gebracht werden können. Da zu dieser Frage auch Experimente zum Differenzierungsverhalten von malignen Zellen unter Therapie mit xenogenen Peptiden und Proteinen vorliegen und deren Toxizität bekanntermaßen äußerst gering ist, lag es nahe, bei Patienten mit diesen Erkrankungen, bei denen alle denkbaren Therapien ohne Erfolg eingesetzt worden sind, erste Therapiestudien mit diesen Substanzen vorzunehmen. Die Dokumentation von vier klinischen Fällen unter Therapie mit xenogenen Peptiden und Proteinen zeigt, daß dieser Ansatz weitere Forschungen in dieser Richtung rechtfertigt.

References

1   Chen, P.-M.; Ho, C.-K.: Immunological Classification of Acute Lymphatic Leukemia; in Haematology and Blood Transfusion, Vol. 28, Modern Trends in Human Leukemia V, pp. 124–125 (Springer, Berlin-Heidelberg 1983).
2   Moore, M. A. S.; Gabrilove, J.; Sheridan, A. P.: Myeloid Leukemic Cell Differentia-

tion Induced by Human Postendotoxin Serum and Vitamin Analogues; in Haematology and Blood Transfusion, Vol. 28, Modern Trends in Human Leukemia V, pp. 327–335 (Springer, Berlin-Heidelberg 1983).
3   Gaedicke, G.: Leukämie und Zelldifferenzierung. Untersuchungen zur Pathophysiologie an der tierexperimentellen Friend-Virus-Leukämie und der akuten lymphatischen Leukämie des Kindesalters. Aktuelle Onkologie, Band 6 (Zuckschwerdt, München 1982).

Dr. med. Helmut Röhrer, Innere Abteilung, Rosmann-Krankenhaus, Zeppelinstraße 37, D-7814 Breisach (FRG)

Immunomodulation:
A New Therapeutical Method
in Cancer Treatment?

Friedrich R. Douwes

Sonnenberg Clinic, Bad Sooden-Allendorf, FRG

Introduction

Immunotherapy was developed during the 1970's as a promising and important new method in the strategy of cancer treatment. Although immunotherapy is still in its infancy and still largely in an empirical stage, there have been adequate, significant clinical results to justify optimism that this therapeutic method will take its place alongside surgery, radiotherapy, and chemotherapy as an alternative way in the future treatment of cancer patients.

The observation that malignant tumors show spontaneous regression in animals as well as in man, or tend to regress after severe bacterial infection after vaccination with special bacterial materials, permits assuming that there is a biological relationship between host and tumor, which finally determines the cause of the cancerous disease [1]. A successful immunological defense against cancer requires a specific antitumor response, if the basic postulation for one or more tumor-associated antigens are accomplished. Specific tumor antigens using modern technologies can be shown in several human tumors, especially in malignant melanoma [2, 3].

Immunobiology and the Basis for Immunotherapy

The scientific basis for immunotherapy is the discovery that most animal and human tumors express cell-surface antigens of different varieties. These are often recognized as foreign or "non-self" by the tumor-bearing host and thus, an immune response is directed against these antigens. Despite the presence of surface antigens as a target for destroying tumor cells, immunodeficiency is a common phenomenon in many types of human malignancies and becomes severe as the cancer progresses [4, 5]. The etiology of this tumor-associated immunosuppression is complex and can be cause for instant insufficiency of immunocompetent host-cells, due to an intrinsic disorder of immunological stem-cells. The increase of suppressor T-cells is caused by a regulation disorder of the T-cells' dependent function, or caused by production of immunosuppressive factors by the tumor itself (fig. 1). In addition, not to be overlooked that conventional cancer therapy including surgery, radiotherapy, and chemotherapy has immunosuppressive effects, which can be long-lasting. Both general immunocompetence and specific tumor immunity appear to be related to the prognosis of the cancer patient [6]. The host mechanisms controlling cancer growth are not fully understood, but monitoring of immunological parameters allows correlation between the clinical course and effectivity of cancer therapy. Recent therapeutical concepts are based on modulation of the biological response by so-called "biological response modifiers" (table I).

As early as the beginning of the 20th century, *Paul Ehrlich* and others suggested there might be a difference between normal and neoplastic cells, which could be fundamental for a vaccination program against

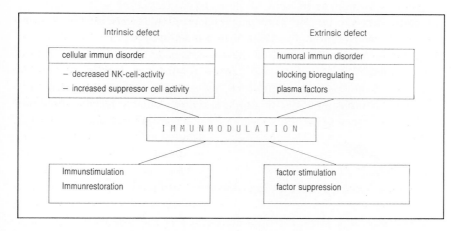

Fig. 1. Cancer and the immune disorder of the host.

Table I. Possibilities of immunomodulation

Immunostimulation
– active unspecific
  BCG, C. Parvum
– active specific
  tumor cells, vaccination

Immunorestoration
Active	– hormones, (thymus factory)
	– chemicals (levamisole)
	– substitution of stimulating regulating mediators (organlysates, interleukins)
	– elimination (suppression) of inhibitory mediators (prostaglandins, immunocomplices)
Adoptive	– transfer of immune competent cells
	– transfer of activated immunocells
	– transfer of cytoplasmic substances for cell activation and regeneration
Passive	– antitumor antibodies
	– tumor necroticizing factor
Preventive	– retinoids

cancer. However, this experimental work was unsuccessful because of the use of non-inbred strains of animals. Thus, the rejection of tumor graft was caused primarily by sensitization to strong histocompatibility antigens. After understanding, tumor immunology became unimportant until fully inbred strains of animals were developed and the rejection of tumor grafts were assumed to be based on immune reactions against specific tumor antigens.

Experiments showing resistance to transplants of chemically induced tumors, finally led to the revival of tumor immunology as an active field for experimental and clinical research. No doubts further that tumor antigens and tumor immunity exist and play an important role in the surveillance of cancer in man. Tumor growth stimulates all major cells of the immune system, which are involved in a differentiated and defined way in the killing of tumor cells. If all these cell types of the immunosystem are really important in human-tumor biology, is not yet clear. Many of these immune mechanisms can be demonstrated in vivo. They can be inhibited or can be stimulated. If soluble tumor antigens or antigen-antibody complexes are shed from the tumor surface, it could be demonstrated that killer cells are inhibited and intolerance against the tumor is induced.

Since generalized immunodeficiency and deficiency of specific host-defense factors are associated with poor prognosis, and since activation of various host-defense components to, or even above the normal level, can result in the killing of tumor cells, it is supposed that, by specific and/or unspecific immunomodulation, an increased antitumor activity can be achieved [1] (table II).

There is sufficient data available to support this interpretation [6–10].

Types of Immunotherapy

In experimental animal and clinical trials, immunotherapy is approached differently (table III). First, the host-immune response must be restored in the partially or completely immune-suppressed cancer patient. This can be done with microbial and synthetic immunostimulants, thymic hormones, organolysates, interleukins, transfer factors, etc. Restoration can also be accomplished by removal, from the plasma, of circulating substances such as soluble tumor antigens, antigen-antibody complexes, etc. which may be suppressing for the immune response. It should be stressed that cytoreduction of tumor mass with surgery, chemotherapy or radiotherapy can also restore the immunity of cancer patients.

However, it should not be forgotten, that these therapeutical methods mainly have an immunosuppressive action. Therefore, a second objective would be to protect the host against or reverse the immunosuppressive effects. A third objective would be to induce an increase in specific tumor immunity or restore tumor immunity, if absent or weak.

The specific modulation of some effective mechanisms aims to increase tumor regression. This leads to an activation of cytotoxic-effector

Table II. Possibilities for immunotherapy

---

(1) Restoration of immunocompetence of immunodeficient patients.
(2) Prevention or reservation of the immunosuppression induced by surgery, radiotherapy, and chemotherapy.
(3) Induction or potentiation of specific tumor immunity.
(4) Modulation of immune response for selected objectives:
    (a) Augmentation of cell-mediated immunity
    (b) augmentation of cytostatic antibodies
    (c) activation of macrophages
    (d) increase of RES clearance
        – reduction of blocking or inhibitory factors
        – reduction of circulating antigen or antigen-antibody complexes

Table III. Principles of immunotherapy

Approach		Agent	Mechanisms	Disease where activity demonstrated
(A)	Systemic Active Non-specific	Immunoadjuvants Immunopotentiators – BCG – C. Parvum – Mycobacteria extracts (MER, CWS) – Pseudomonas vaccine – Mixed bacterial vaccines – Sulfated organolysates Immunorestorative – Levamisole	Increased general competence, increased reticuloendothelial activity Restored immunocompetence	Melanoma Lung cancer Breast cancer Colon cancer Myeloma Head and neck cancer
(B)	Systemic Active Specific	Unmodified tumor cells Modified tumor cells Tumor antigens	Increased tumor Specific cellular and/or humoral immune response	Melanoma Lung cancer Hypernephroma
(C)	Local	Haptens (DNCB) Virus BCG Bacterial antigens	Active macrophages Destroy tumor by Bystander effect, induce tumor immunity	Skin cancer Melanoma Breast cancer
(D)	Adoptive	Lymphocytes Immune RNA Transfer factor	Transfer specific Immunity	Melanoma Myeloma
(E)	Mediators and hormonal	Lymphokines Interferon Thymic hormones Tumor Necrosis factor		Osteogenic Sarcoma Myeloma
(F)	Passive	Allogenic or xenogenic antibody Plasmapheresis	Transfer cytotoxic or opsonizing antibody Couple with chemotherapy Removal of tumor antigen and antigen-antibody complexes	

cells and an increase in the clearing capacity of the RES by elimination of blocking factors and immunocomplexes. This can be achieved by active and specific immunization. Immunopotentiators can also enable the patient to respond to otherwise poorly immunogeneic antigens.

The possible principles of immunotherapy are demonstrated in table III. There are six major categories of immunotherapy:
(1) Systemic active non-specific immunotherapy;
(2) systemic active specific immunotherapy;
(3) adoptive immunotherapy;
(4) passive immunotherapy;
(5) local immunotherapy;
(6) combination immunotherapy.

Immunostimulation and/or Immunorestoration

Systemic active non-specific immunotherapy includes the use of adjuvants usually of microbial origin. The active non-specific immunomodulation influences different parts of the lymphoid and non-lymphoid cell population and their mediators.

BCG – an Active Non-Specific Immunostimulant

The substance used frequently as immunostimulator is BCG (bacillus Calmette-Guerin). But beside this, non-viable fractions of mycobacteria such as methanol extraction residence of BCG (MER) and other cell-wall fractions (cell-wall skeleton) are used, as well as Corynebacterium Parvum or granulosum-lipopolysaccharides, including pseudomonas vaccine, bacterial toxins, Pertussis vaccine, vaccine virus, polysaccharide extracts of fungi, synthetic adjuvants, including double-stranged RNA, pyrancopolymer, and poly-I-poly-C are further modalities. The BCG treatment in cancer patients, especially patients with melanoma, has been studied in several clinical trials. The results of a local or systemic BCG treatment are still controversial [7, 8, 11]. However, the clinical data obtained through the development of BCG immunotherapy has led to the basic principles of active non-specific immunotherapy. Only the administration route has been different. The optimal route of administration depends both on the type of agent given, as well as on the tumor site. In some experimental models, direct contact between the immunotherapeutic reagent and the tumor cells is required for maximal antitumor activity. In such experimental or clinical models, macrophage-killing is important. Macrophage-activation probably occurs via the release of lymphokines from antigenically stimulated T-cells. Similar mechanisms may play a role in some of the clinical experiments with regional or systemic immunotherapy. The work with BCG for metastatic intradermal melanoma modules showed that it is necessary to inoculate BCG directly into the tumor. For other types of cancer, the local application was modified orally for gastrointestinal neoplasms, intrapleurally for lung cancer, intraperitoneally for a tumor arising in the peritoneal cavity, intradermally for residual malignant melanoma. In animal models, results of this kind of immunotherapy were better than those obtained in man [7].

The active non-specific immunomodulation increases cell-mediated immunity, macrophage activity and antibody-mediated cytotoxicity, as well as natural killer-cell activity (NK-cell activity) [8]. It leads to a reduction of suppressor cells and induces an increased release of pluripotent hematopoetic stem-cells and lymphoid cells. Complement also can be activated [10].

At present, it is not exactly known which is the best route for the optimal dose and application of BCG. One of the basic principles of clinical pharmacological studies, is the establishment of the dose-response. This should be done in careful clinical studies according to an evaluation model modified from the principles for chemotherapy. It is surprising that careful dose-response studies have rarely been carried out in clinical immunotherapy studies. Too low or either excessive doses are frequently ineffective or even lead to enhancement of tumor growth because of immunoparalysis. These controversial results of clinical studies brought clinical use of immunotherapy.

Nevertheless, there are several measurable effects of immunotherapy. The regional effect of immunotherapy has been shown in studies in melanoma and early lung cancer. BCG applied in melanoma by scarification into the upper and lower extremities, prolonged the postoperative disease-free interval and survival of the patients. Additionally, it was demonstrated that BCG could potentiate the antitumor effect of cytostatic agents such as DTIC. The immunomodulation induced by C. Parvum is similar to that of BCG. In particular, the activation of macrophages is responsible for the antitumor activity. The same can be assumed for the other non-specific immunomodulators. Apart from the effect on T-cells and macrophages, there also is a reduction in suppressor-cell activity so that cytotoxic systems can be activated.

Most of the synthetic chemicals of different structures used as biological response modifiers are, currently, in experimental use or in phase-I studies. The common property of these substances is obviously the increase in interferon production and activation of NK-cells. Since there is some evidence that interferon may have direct antitumor activity or may have adjuvant activity in antitumor systems, these compounds are of great interest. A number of these materials such as levamisole or poly IC have also immunoprophylactic activity and protect animals from subsequent viral and chemical carcinogens and from the subsequent implantation of transplantable tumors. In addition, these substances can prolong remissions in humans induced by chemotherapy. The immunological characteristics of these materials are, in addition to the introduction of interferon, the augmentation of antibody production and the activation of macrophages.

Levamisole – Unique Immunological Reagent

Most important among these reagents is levamisole, which has already shown some activity in immunotherapy of animal and human tumors. Negative skin tests to recall antigens will change to positive and impaired phagocytosis will disappear [11]. Although the therapy is not free of side-effects, patients with lung cancer show a prolonged disease-free interval after resection. The survival time of these patients is significantly prolonged [12]. Originally, levamisole had been used for years as an antihelmintic agent. The first observation in man was an increased tuberculin and DNCB reactivity among immunodeficient persons after the administration of levamisole. No effect could be seen on the immunological reactivity of normal persons. Several clinical studies have been made in a number of tumors such as lymphomas, breast cancer, lung cancer, etc. The effectivity of this type of immunotherapy has already been reported for breast cancer and lung cancer. In a study on lung cancer stage I, patients received 150 mg levamisole in three divided doses per day for three consecutive days; repeated every 2 weeks starting 3 days prior to surgery. There was significant improvement in survival.

In another study, patients with inoperable stage-III breast cancer were placed on a similar regime after being free from disease after radiotherapy. There was significant prolongation in the levamisole group of both the mediane disease-free interval from 9 to 25 months and the survival time. Finally, in advanced breast cancer, levamisole added to chemotherapy-prolonged remission time and survival time, compared to the use of chemotherapy alone.

Thymus Hormones

A working thymus gland is necessary for the development and regulation of cell-mediated immunity. Those functions under thymic control include primary transplant and tumor immunity, as well as viral, mycobacterial, fungal, and protozoal immunity. Thymic-dependent lymphocyte populations are directly involved in the function of T-cells and the

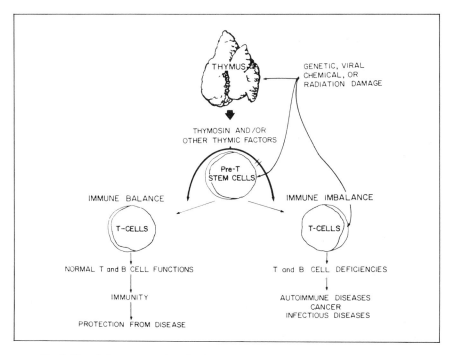

Fig. 2. Thymic hormone action. Lack of normal hormone-production results in immune imbalance, leading to disease.

subpopulations, e.g., helper-cells, suppressor-cells, effector-cells, etc. As shown in figure 2, it is evident that a part of a thymic-dependent process occurs via hormonal mechanisms. The thymus produces thymosin and a wide spectrum of other thymic hormones, e.g. thymopoetin. Studies show that a deficiency in thymic factors results in an immune imbalance, leading to a number of serious diseases, including autoimmune diseases, cancer and infections; moreover, studies made with thymosin fraction V showed that it corrects some immunodeficiencies in man. In the partially purified fraction V are a number of biologically active peptides, which contribute to its biological activity [13].

The main functions of thymic peptides are listed in table IV.

Restoration of the thymic-dependent immunological capacity and an increase in specific cell-mediated immunity is most important.

The systemic application could, therefore, be a promising therapeutical method in several, very different diseases. *Goldstein et al.* [14, 15] showed that patients with immunodeficiency, autoimmunodiseases, cancer, allergic diseases, acute and chronical infections, increase their number of immunocompetent T-cells by application of thymic hormones. In a clinical study with 277 patients with head- and neck-, gastrointestinal- and skin cancer, patients with thymosin in combination with radio- or chemotherapy showed increases of survival time. Patients with oat-cell cancer who, in addition to conventional chemotherapy, received thymosin fraction V, showed a significantly increased survival time in com-

Table IV. The main functions of thymic peptides

Experimental	Clinic
Increase of survival time of thymectomized animals	Increase of T-lymphocytes in thymus hypoplasie
Restoration of thymus-dependent immunity of nude mice	Increase of survival in gastro-intestinal, skin, lung, head and neck cancer
Increase of resistence to spontaneous tumors in mice	Increase of T-cells in primary immunodeficiencies, autoimmuno-diseases, allergies, infections
Increase of specific immunity	(viral, bacterial)

parison to the control group especially if they had low T-cells before therapy. We have had experience with NeyThymun® (vitOrgan, Ostfildern/Stuttgart). This thymus preparation contains a mixture of cytoplasmic peptides and proteins, which are obtained by a special extraction procedure (DB 8 No. 1090821, 2819131). By this procedure, extractions are either from fetal or juvenile calf thymus. In comparison to other thymus preparations, NeyThymun® shows some exceptional advantages:

(1) Differentiation of fetal and juvenile thymus: NeyThymun F (fetal thymus), NeyThymun K (juvenile thymus);

(2) application possible from 10^{-12} g protein/ml to mg protein/ml, according to tolerance;

(3) free of toxic preserves;

(4) standardized by bioactivity in human-cell culture.

The originally postulated difference by *Theurer* between fetal and adult thymus function was confirmed by several other authors. The fetal thymus serves in the protection of gravidity; the immunosystem of the mother is being suppressed by fetal thymus to protect the fetus. *Zaplicki et al.* [16] showed a decrease of leucocytes and gammaglobulines after application of fetal thymus extract, after a lethal dose of X-ray to mice in the fetal-thymus group, the survival time was significantly reduced. Therefore, the indications for fetal thymus at the moment (NeyThymun F) are hyperegic diseases, allergies, and rheumatic diseases.

The juvenile thymus (NeyThymun K) increases the self-protection against microbial invasion and non-self antigens. The detection of foreign antigens is of importance in the control of bacterial, viral diseases as well as in the control of malignant tumor cells. Therefore, the reagent is used in diseases with impaired immunity especially to stimulate cell immunity and RES (see figure 3).

Thymus Factors Combined with Factors from Different Organs (NeyTumorin®)

Recent studies showed that factors extracted from cytoplasma of other organs such as liver, placenta, and bone-marrow, contain a number of highly bioactive factors, which are able to regulate very specific properties of cells. The organs seem regulated by a number of signal peptides which coordinate the intracellular correspondence in a healthy body. These factors are obviously phylogenetically older than the highly differenti-

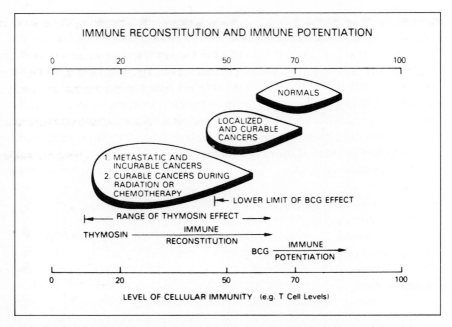

Fig. 3. Schematic representation of the differing effects on immune reactivity of thymosin and BCG.

ated immunosystem. The preparation used in cancer therapy is NeyTumorin®. We have had our own experience in a number of malignant diseases and systemic disease such as myeloma. In 16 cases of advanced multiple myeloma, stage III, resistant to any kind of chemotherapy, we could see 10 complete and partial remissions under monotherapy with NeyTumorin®. Results of this study are published elsewhere [17].

Tumors of the lymphatic system are associated with several defects of cellular and humoral immunity [18, 19]. As already mentioned, treatment of these diseases with radiation and alkylating agents further reduce immunocompetence favoring the occurrence of severe infections' disease. Since thymus extracts have reconstituting capacity, both in vitro and in vivo, on depressed immunological parameters in various immunodeficiencies, we also tested the effects of a NeyTumorin® application in patients with lymphoproliferative disorders. Fourteen patients, aged from 32 to 64 years, were included in this study. Five suffered from Hodgkin's disease, six from Non-Hodgkin's disease and 3 from chronic lymphatic leukemia. In all patients, T-cells were less than 50% of the Ficoll-separated mononuclear cells, all patients had received alkylating agents and/or radiation, but at the time of treatment with NeyTumorin®, the patients only received this substance in a dosage of 0.5 mg/kg B.W. given i.v. 3 times weekly for at least 3 months. Comparison of the total WBC-count, absolute number and percentage of T- and B-cells, before and after NeyTumorin®, showed that patients with Hodgkin's disease and Non-Hodkin's disease had a significant ($p < 0.05$) increase in the absolute number and percentage of T-cells and no significant change of the white blood count and of the total number and percentage of B-cells (table V + VI). The three patients with chronic lymphatic leukemia were analyzed separately and in these patients

Table V. WBC and absolute number and percentage of T- and B-cells (ISD) after treatment with NeyTumorin-Sol®

Diagnosis	Pat. (n)	WBC	After treatment with NeyTumorin-Sol®				
			Total T-cells	T-cells in %	Total B-cells	B-cells in %	Significance (p)
Normal Controls	20	7.673 ±1.921	–	–	–	–	–
Hodgkin's disease	5		3.583,3 ± 529	46,7 ± 6,9	913,0 ±268,6	11,9 ± 3,5	$p \leq 0,05$ for T-cells NS for B-cells
Non-Hodgkin's disease	6	6.751 ±1.912	3.017,7 ± 803.4	44,7 ±11,9	837,1 ±310,5	12,4 ± 4,6	$p \leq 0,05$ for T-cells NS for B-cells
Chronic lymphatic leukemia	3	32.312 ±5.676	9.402,8 ±3.748	29,10 ±11,60	19.710,0 ± 3.005,0	61,0 ± 9,3	NS for T-cells NS for B-cells

Table VI. WBC and absolute number and percentage of T- and B-cells (ISD) before treatment with NeyTumorin-Sol®

Diagnosis	Pat. (n)	WBC	Before treatment with NeyTumorin-Sol®			
			Total T-cells	T-cells in %	Total B-cells	B-cells in %
Normal Controls	20	6,912 ±1,715	4,572.2 ±1,320.2	66.15 ±19.10	1,016.1 ± 518	14.7 ± 7.5
Hodkin's disease	5	6,115 ±1,495	1,473 ± 669.6	24.10 ±10.95	611.5 ± 171.0	10.0 ± 2.8
Non-Hodgkin's Disease	6	6,710 ±1,512	1,952 ±9,538	29.10 ±14.20	899 ± 376	13.4 ± 5.6
Chronic lymphatic leukemia	3	37,585 ±6,453	6,502 ±4,172	17.30 ±11.10	28,038.4 ±3,382.7	74.6 ± 9.0

it could be demonstrated that T-cell level increased, but no significance could be seen on the B-cell level, although the patients showed an improvement clinically.

The results are in accordance with results observed in patients suffering from melanoma and gastrointestinal tumors who had been treated with thymic factors. This indicates that sulfatated organolysates have, besides cell proliferation inhibiting effects, a restoring potency on cellular immunity in patients with lymphoproliferative diseases. This and other results, obtained in preliminary clinical trials, justifies checking the response of this reagent in cancer patients.

Interleukins

The normal functions of different cell compartments of the immunosystem (T-cell with its subfractions, B-cells, monocytes, macrophages, etc.) are regulated by soluble mediators such as lymphokines, recently called "interleukins". The aging of the immunosystem correlates with a decrease in the T-cell function and this is dependent on the ability to produce interleukin-2. Using interleukin-2 extracted from in-vitro cultures, effector T-cell function can be reconstituted. Because interleukin can be inactivated easily, inclusion in liposomes is preferred for application.

Active and Passive Immunotherapy

Usually, there are two ways for a specific immunotherapy:
(1) Active vaccination with specifically treated inactivated autologous tumor cells, or homologous tumor cells to induce a cellular and humoral antitumor reaction;
(2) passive immunotherapy with antibodies against tumor antigens (specific immunoglobulins).

The active specific immunotherapy has been used successfully since 1969 in acute lymphoblastic leukemia. Since then, survival studies have been made showing the effectivity of such programs. The dose of these substances for the active specific immunotherapy has to be optimized to prevent one occasionally observed side-effect – the tumor enhancement!

Summary

Cancer treatment in its present stage seems to be limited, response rates and survival times are no longer changing. This fact should assist us in regarding the conventional concepts, but also inspire us in investigating new, proven ways, using the vast, new knowledge which has recently been gained about the interaction between the tumor and its host. The systematical proof of modulation of tumor-host inhibition is necessary.

Zusammenfassung

Die Tumorbehandlung in ihrem gegenwärtigen Zustand scheint sich in einem no change der Ansprechraten und Überlebenszeiten zu befinden. Aus diesem Grund müssen neue Wege der Tumortherapie eröffnet werden, insbesondere deshalb, da inzwischen umfangreiche Kenntnisse über die Interaktion zwischen Tumor und Wirt bestehen. Eine systematische Untersuchung zur Modulation des Immunsystems bei Tumorpatienten erscheint notwendig.

References

1. Nauts, H.; Fowler, G. A.; Bogatho, F. H.: Acta med. scand. (Suppl.) *145:* 1 (1953).
2. Old, L.: J. Cancer Res. *41:* 361 (1981).
3. Mutzner, P. A.: Schweiz. med. Wschr. *111:* 1322 (1981).
4. Vaage, J.; Doroshow, J. H.; du Bois, T. T.: Cancer Res. *34:* 129 (1974).
5. Douwes, F. R.: Immundiagnostik maligner Erkrankungen (Fischer, Heidelberg 1979).
6. Adler, A.; Stein, J. A.; Ben-Efraim, S.: Cancer *45:* 2074 (1970).
7. Baldwin, R. W.; Hoojer, D. G.; Pimm, M. V.: Ann. N.Y. Acad. Sci. *277:* 124 (1976).
8. Pioch, Y.; Gerber, M.; Serrou, B.: Cancer Immunol. Immunother. *7:* 181 (1979).
9. Uchida, A.; Hoshino, A. T.: Cancer *45:* 476 (1980).
10. Pehamberger, H.; Holubar, U.; Knapp, W.: Cancer *46:* 1135 (1980).
11. Lederer, E.: Synthetic immunostimulants derived from the bacterial well. J. med. Chem. *23:* 819 (1980).
12. Amery, W. K.: Double blind levamisole trial in resectable lung cancer. Ann. N. Y. Acad. Sci.: *277:* 260–268 (1976).
13. Goldstein, A. L. et al.: Thymosin: chemistry, biology and clinical applications; in van Bekkum (ed.), The Biological Activity of Thymic Hormones, pp. 173–197 (Kooyber Sci. Publ., Rotterdam 1975).
14. Goldstein, A. L. et al.: Use of Thymosin in the treatment of primary immunodeficiency disease and cancer. Med. Clins N. Am. *60:* 591–606 (1976).
15. Goldstein, A. L. et al.: First clinical trial with Thymosin. Reconstitution of T-cells in patients with cellular immunodeficiency diseases. Transplant. Proc. *7:* 681–686 (1975).
16. Zaplicki, J. et al.: Thymus *3:* 143–151 (1981).
17. Douwes, F. R.: Therapie des fortgeschrittenen Plasmozytoms mit immunbiologischen Substanzen. (Publ. in Vorbereitung).
18. Chirigos, M. A. et al.: Modulation of immunity in cancer by immune modifiers. (Raven Press, New York 1981).
19. Tursz, T. et al.: Low natural killer cell activity in patients with malignant lymphoma. Cancer *50:* 2333 (1982).

Prof. Dr. med. Friedrich A. Douwes, Ärztl. Direktor der Sonnenberg-Klinik, Hardtstr. 13, D-3437 Bad Sooden-Allendorf (FRG)

Cytobiological-Cytostatic Combination Therapy
A New Approach in Medicamental Oncotherapy

B. Kisseler[1], Th. Stiefel[2]

[1] Central Radiological Department of Böblingen District Hospital, FRG
[2] Stuttgart, FRG

During the past ten years, chemotherapy has progressed in clear, if not overwhelming, achievements. These have largely been due to the insight that the majority of malignant diseases call for joint collaboration of surgeons, radiotherapists, and internal specialists in the elaboration of a therapeutic concept. Further, progress has been attained by the introduction of new cytostatic agents, through modifications in the application of known cytostatic agents and through the improvement of supportive measures, which are the precondition for the application of an intensive cytostatic therapy. Nevertheless, the selectivity and the therapeutic range of the cytostatic agents currently available are still unsatisfactory. Toxic side-effects, frequently of a specific nature in certain organs, restrict their application. As an example, mention is made of the cardiotoxic effects of adriamycin [1], the nephrotoxic effects of mitomycin C [2, 3] or the bone-marrow depressions and nausea which occur, to a greater or lesser extent, with all cytostatic agents (alkaloids, antibiotics, alkylating agents, antimetabolites).

The subjectively poor, general state of health of patients under chemotherapy often leads to a discontinuation of the therapy in clinical practice. In addition, a higher-dosage chemotherapy cannot be justified for patients in such a poor general condition. For this reason, for some time now, antiemetics, vitamin preparations or blood derivatives have been used in addition to the administration of cytostatic agents. However, the substances used up till now, under these aspects, do not dispose per se of any antitumoral efficacy. In the search for antiemetic substances with an antitumoral action, we have come across biological active agents in the form of exogenic peptides and proteins (NeyTumorin®, vitOrgan, Ostfildern, FRG) which are isolated from healthy and non-malignant tissue of various species. The antitumoral efficacy of these substances has been demonstrated in comprehensive in vitro and in vivo experiments [4–12]. The intended effect of these substances has also been shown clinically, as a monotherapy and in combination with chemotherapy. A noteworthy feature was the good general state of health of patients under this therapy. Consequently, for treatment of extensive inoperable tumors or with recurrent carcinomas or such with a generalized formation of metastases, a synergistic action in the sense of an enhanced antitumoral efficacy with improved tolerance would be expected for the combination of such substances with potent cytostatic agents.

For this reason, as an initial step, various combinations of cytostatic agents with NeyTumorin were studied. Within the framework of comprehensive investigations, it was found that mitomycin C [2, 3], which has been used by us in many tumor cases but has the disadvantage of having considerable side-effects, can be combined with NeyTumorin®. To begin with, suitable incubation conditions were established under the aspect of a solution of this combination for use as an intravenous infusion, with the aim of obtaining clear solutions without a precipitate. In a further step, the molecular interactions between the combined substances were studied and finally, the antitumoral efficacy of the individual substances and combinations were measured in human tumor-cell cultures or diploid fibroblast cultures. The ^{3}H-thymidine incorporation rate was used as a marker for the DNA-synthesis of the cell-lines studied. In addition, the proportion of live cells to dead cells after the treatment was determined with the aid of trypan-blue staining. We have also found the combination of mitomycin C with adriamycin to be effective. However, the study of the combination capacity with NeyTumorin®-Sol revealed the occurrence of non-specific precipitates. Finally, it was established in large-scale mass examinations that adriamycin can be combined with a hemoderivative of calf's blood. This combination was alo studied with respect to the antitumoral efficacy of individual substances and combinations in tumor- and normal-cell cultures. The findings from these experiments then led to the elaboration of a therapy plan which, in the meantime, has been used for several patients with a noteworthy degree of success.

Experimental Part

Combination of Mitomycin and NeyTumorin®

Incubation

NeyTumorin® and mitomycin were dissolved in physiological saline solution and incubated in a ratio of 15 mg NeyTumorin® to 2 ml mitomycin for 12 h at 37° C, stirring being maintained. Non-incubated mixtures of the same combination, which were mixed immediately before use in the cell-culture, and the individual preparations served as reference samples.

HPLC-Separations

The HPLC-separations were carried out under the following conditions: separating column; gel-type: G 3000 SW (LKB 2135/60); column dimensions: 600 × 7.5 mm; preliminary column: gel-type Ultro-Pak-TSK GSWP: column dimensions: 75 × 7.5 mm; flow-rate: 50 µl/min. After passing through the separating column, the fractions were measured on-line in a flow-photometer at 280 nm. The sample volume was 250 µl. A buffer solution with 0.1 M Na_2HPO_4, 0.1 M NaCl and pH 7.2 was used as the eluate buffer. The separated fractions were collected with a fraction-collector so that their mitomycin content could be studied and their antitumoral efficacy in cell-cultures tested.

Mitomycin Determination

The exposure of the UV-vis spectrum of mitomycin revealed an absorption maximum at 218 and 365 nm. Since the peptide and protein fractions of NeyTumorin® do not absorb at wavelength 365 nm, it was possible to determine the mitomycin content of the fractions obtained after separation of the NeyTumorin®/mitomycin mixture by measuring the absorption at 365 nm. It was also demonstrated that the characteristic absorption maximum of mitomycin has not changed after incubation with NeyTumorin®.

Cell-Culture
DNA-Synthesis

Human tumor cells (Hep 2, Wish) and diploid fibroblasts (FH 86) were used as test-cells. All the cell-cultures passed through a test-cycle consisting of a growth-cycle and a test-phase. The test-substances were used in the test-phase.

The cells, which had been kept under liquid nitrogen, were thawed out at 37° C and incubated in culture flasks with the minimal essential medium (Gibco Europe, Karlsruhe, FRG) with additions of Hanks' salts (MEM-H), penicillin, streptomycin, neomycin, and 5–10% foetal calf serum. After two days, the cells were visually counted after trypsinization and again seeded. Following daily medium changes, the cell-count was repeated on the 4th day and 1×10^5 cells per 1 ml MEM-H + 10% FKS seeded out on special cell-plates, each containing 30 single-cell cultures of 1×10^5 cells.

Test Phase

The preparation was added after 40 h, 50 µl preparation per flask being added in the present tests. Six parallel samples were prepared per test for each type of preparation and concentration. Incubation was limited to 8 h. 4 h before the end of this exposure time, 50 µl methyl-^{3}H-thymidine solution (0.35 µCi) werde added by pipette in each case. After further 4 h, the non-incorporated ^{3}H-thymidine was removed by washing four times with 1 ml 2% perchloric acid each time and hydrolyzing the cells with 1 N HCl at 70° C for 1 h. The hydrolysate was transferred to scintillation vials and mixed with cocktail (Aqua Sol 2, NEN). The activity measurement was carried out in a Beckman LS-100 liquid scintillation counter. Relative standard deviations of $s_r = \pm 10\%$ were determined as statistical quality coefficients of the total test system.

Combination of Hemoderivative with Adriamycin

Incubation

One vial adriamycin (10 mg) was suspended in 10 ml hemoderivative and incubated for 12 h at 37° C. Non-incubated samples and the individual substances served as reference samples.

Cell-Culture

The measurements of the DNA-synthesis with this combination were carried out as described under the combination of mitomycin/NeyTumorin®.

Trypan-Blue Staining

This staining test provides information on the vitality of cells, based on the fact that live cells do not absorb stains such as trypan-blue, whereas with dead cells both the cytoplasm and the nucleus can be stained [17–20].

The cells were processed as described above and, instead of the radioactive labeling, stained with trypan-blue (3.3'-[(3.3'-dimethyl-4.4' biphenylene)-bis) azo)]-bis (5-amino-4-hydroxy-2.7-naphthalene disulfonic acid)-tetrasodium salt). A basic condition is that the cells are washed protein-free before the staining to prevent non-specific staining at proteins. The concentration of trypan-blue was 0.18%. The test-solution was mixed with a pipette and allowed to stand for 4 to 5 min at room temperature. It is then once more vigorously mixed and a counting chamber (Fuchs/Rosenthal, FRG) filled with a few drops. The cell-count is started and the percentage of live cells calculated with the equation

$$\% \text{ live cells} = \frac{\text{unstained cells}}{\text{stained and unstained cells}} \times 100$$

All cells whose nucleus or cytoplasm displays a blue stain are considered to be stained and thus dead, or damaged. Even those cells which display only a pale blue staining of the cell-nucleus are designated as dead.

Results

Mitomycin/NeyTumorin® System

In a test solution, the DNA-synthesis of tumor cells and normal cells was measured after being treated for 8 h with NeyTumorin or mitomycin, the NeyTumorin® + mitomycin mixture and the incubated product of NeyTumorin® + mitomycin. A comparison was made of the effect with a pure physiological saline sample and of an albumin sample with the same protein content as NeyTumorin®. Figure 1 shows that in comparison with the control-culture treated with physiological saline solution, or the culture treated with bovine albumin, both NeyTumorin® + mitomycin and NeyTumorin® incubated with mitomycin are extremely effective. The effect of pure mitomycin, especially in the group treated with incubated NeyTumorin®/mitomycin, is even surpassed in the tumor cells. The mixing of NeyTumorin® and mitomycin without incubation, led to comparable inhibitory results with the normal-cell culture, whereas a weaker effect was observed with tumor cells. NeyTumorin® alone tends to stimulate the normal cells while tumor cells are inhibited. Whereby, note that this biological substance causes a far lower degree of mitotic inhibition than mitomycin.

The separation of the incubated sample by HPLC revealed that mitomycin is largely bound to 2 low-molecular weight fractions of NeyTumorin® in the molecular weight range between about 1000 and 10 000. Whether this linkage is covalent or came about through hydrogen ring-structures or Van-der-Waals forces cannot be determined from these studies. It is to be noted, however, that this linkage does not break up during the HPLC-separation. Due to these results, the various peaks were

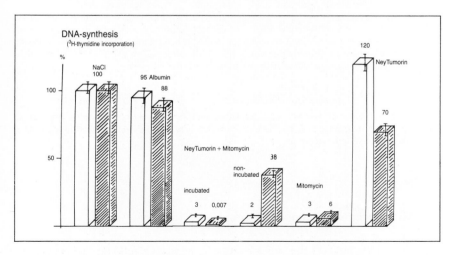

Fig. 1. Behavior of DNA-synthesis (^{3}H-thymidine incorporation rate) after 8-h incubation with NeyTumorin alone, mitomycin alone, NeyTumorin® + mitomycin incubated and non-incubated, with human tumor and normal cells. The effects were compared with bovine albumin and physiological saline solution as control-cultures.

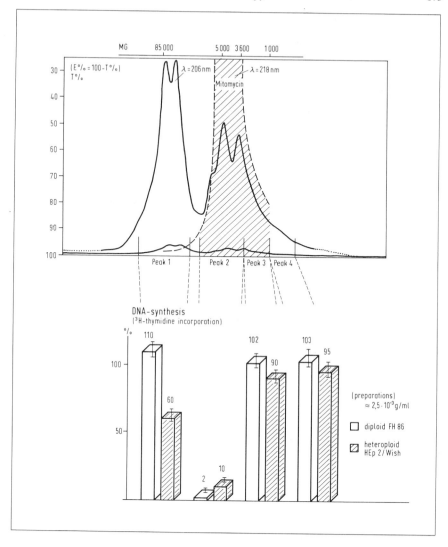

Fig. 2. Separation of incubated NeyTumorin® and mitomycin by HPLC. Mitomycin is again found in the range of two NeyTumorin peaks (peak 2 and peak 3). The result of the DNA-synthesis measurements (^{3}H-thymidine incorporation rate) after 8-h incubation with the individual fractions is given below. It is clear that only peak 2, as the peak containing mitomycin has an effect. The specific inhibitory effect of peak 1 on tumor cells with its higher molecular weight is considerable.

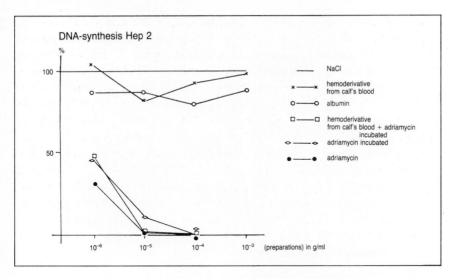

Fig. 3. Dose-effect relation of hemoderivative and adriamycin and hemoderivative and adriamycin incubated and non-incubated in tumor cells.

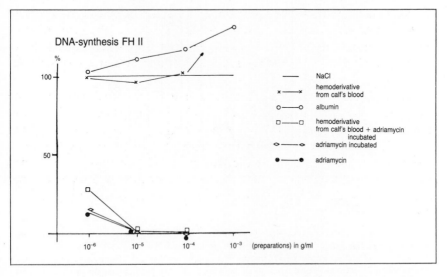

Fig. 4. Dose-effect relation of hemoderivative, adriamycin and the combination hemoderivative + adriamycin, incubated and non-incubated with diploid fibroblasts.

collected separately and their mitotic inhibitory effect studied in diploid and heteroploid cell-cultures. The absorption diagram of the incubated NeyTumorin®/mitomycin sample is shown in figure 2. The broken line represents the results obtained from the mitomycin-absorption measurements. Below this, the results shown were obtained after the study of individual peaks in the cell-culture with respect to the DNA-synthesis rate. Apparently, the mitomycin (MW 334) is bound to low-molecular weight peptide fractions of NeyTumorin®. The nature of these linkages cannot be determined from our results. The study of individual peaks in the cell-culture revealed a clear, mitotic inhibition effect of peak 2, whereas peak 3, which was likewise shown to contain mitomycin, does not display any antimitotic activity. The selective inhibitory effect of the high-molecular weight NeyTumorin® fractions observed in tumor cells is noteworthy.

The Hemoderivate/Adriamycin System

As with the NeyTumorin®/mitomycin system, a study was made of the influence of this combination on the DNA-synthesis of tumor and normal cells. In addition, the dose-dependence of the effects was investigated. Notice that hemoderivative and bovine albumin behave in a comparable manner with respect to the DNA-synthesis. However, the incubation of hemoderivative with adriblastin leads, even at low concentrations, to an extreme inhibition of the DNA-synthesis in normal and tumor cells (s. figs. 3, 4).

The results of the trypan-blue staining after treatment with the same substances show the same relations, but even more clearly. Whereas in the control, the hemoderivative group and the albumin group, it was found that there were approximately the same percentages of live cells; practically all the cells were destroyed in all cell-cultures to which adriamycin had been added. A difference between pure adriamycin and adriamycin incubated with hemoderivative was not apparent (fig. 5).

Clinical Part

To begin with, on the basis of the experimental results, tolerance studies were carried out in animal experiments. After observing a clearly

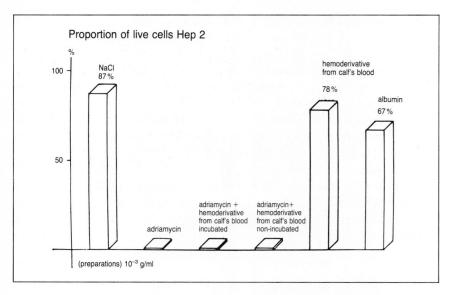

Fig. 5. Results of trypan-blue staining as proportion of live cells of human tumor-cell lines HEP 2 after 8-h incubation with the test preparations. No differences were apparent between adriamycin and the combinations adriamycin + hemoderivative, incubated and non-incubated.

better tolerance of the same doses of cytostatic agents, when administered with appropriate xenogeneic peptides, the first clinical tolerance studies were carried out. It was found that the tolerance of the same doses of cytostatic agents has increased not only subjectively but also objectively, as demonstrated by hematological blood parameters. Even in cases which appeared hopeless, noteworthy remission and response rates were still found.

The fundamental advantage of this cytostatic combination therapy with xenogenic peptides is that it is not associated with considerable subjective impairment to the general condition with side-effects such as vomiting, giddiness, loss of hair, etc. lasting, more or less, for a long period. In addition, the relatively constant hematological laboratory parameters point to an insignificant effect on the myeloproliferative system. Furthermore, the biological cytostatic combination therapy has the advantage that the immune system is not suppressed to the extent observed when cytostatic agents alone are used.

The treatment plan shown in table I was employed for the cases

Table I. Therapy plan

Day 1	20 mg mitomycin are incubated for 12 h at 37° C with 300 mg NeyTumorin-Sol. For the infusion, this mixture is added to 250 ml actovegin 20% per infusion and slow intravenous infusion carried out. 20 mg adriblastin, incubated for 12 h at 37° C with 20 ml actovegin 20%, are then added to 50 ml physiological saline solution and intravenous infusion is likewise carried out.
Day 5	Intravenous infusion of 300 mg NeyTumorin®-Sol in 1000 ml physiological saline solution.
Day 10	20 mg adriblastin, incubated for 12 h at 37° C with 20 ml actovegin 20%, are added to 50 ml physiological saline solution and intravenous infusion carried out.
Day 15	Intravenous infusion of 300 mg NeyTumorin-Sol in 1000 ml physiological saline solution.

Pause of 4 weeks.

Continue therapy until normalization of clinical parameters.

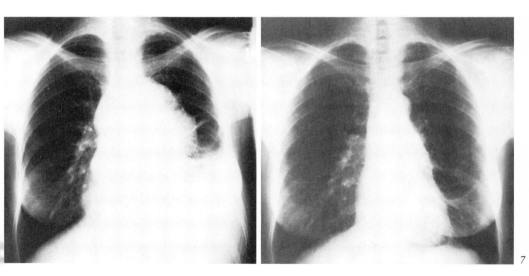

Fig. 6. B.-E. K. Malignant thymoma affecting the left root of the lung, pericardium, mediastinum, and aorta. Left pleura effusion.

Fig. 7. B.-E. K. Follow-up of the malignant thymoma under cytobiological-cytostatic combination therapy with NeyTumorin®, mitomycin, and adriblastin. Complete regression of malignant thymoma (here also see fig. 8).

Table II. Laboratory values

Date	31-5-83 – 24-1-84	13-6-83 – 24-1-84	19-6-83 – 7-2-84	9-5-83 – 11-12-83	2-6-83 – 19-1-84	27-1-83 – 12-12-83
Case	(1) B. E. K. ♀	(2) V. T. ♀	(3) K. G. ♂	(4) J. W. ♀	(5) M. S. ♀	(6) A. S. ♀
Blood picture						
Hemoglobin	10.6 12.0	14.8 13.6	15.1 13.4	16.3 14.7	10.7 12.6	13.5 10.8
Hematocrit	40.0 35.3	42.5 39.7	44.2 40.2	48.5 43.0	32.0 37.2	40.0 31.7
Thrombocytes	196 118	158 152	267 382	185 136	150 190	244 13
Leukocytes	4.5 4.8	2.8 3.3	10.1 11.4	5.6 7.2	2.0 4.4	4.8 4.4
Neutrophils	72 64	54 60	72 62	61 74	66 62	66 64
Monocytes	– 1	1 –	– –	– –	– 5	6 11
Eosinophils	2 2	2 2	2 4	8 2	2 6	3 –
BSR	21/50 24/50	10/26 55/67	66/144 40/87	5/17 5/18	16/37 20/45	45/94 65/125
Erythrocytes	2.97 3.29	4.28 3.96	3.96 374	4.91 4.51	3.65 3.99	4.39 3.46
Immunoglobulins in serum						
IgA in serum	168 220	139 225	779 627	289 302	205 300	276 328
IgG in serum	1064 1078	860 786	1548 1751	1337 1581	1548 2013	1900 1092
IgM in serum	262 227	133 135	108 84	125 127	121 132	100 64
Albumin in serum	4222 3973	4064 4009	3661 3323	5212 4348	4617 4421	4464 2409
C-reactive protein	Pos. 0.7	0.7 0.7	0.7 15.3	0.7 0.7	0.7 0.7	Pos. 17.6
Alpha-2 macroglobulin	187 169	259 226	257 195	214 160	195 281	163 132
Alpha-1 glycoprotein	68 65	69 150	107 163	61 64	61 73	107 148
Alpha-2 haptoglobulins	85 58	196 347	544 727	169 141	76 100	290 248
Coeruloplasmin	32 28	26 34	47 43	56 44	35 31	43 45

IgA, IgG, IgM – turbidimetric,
albumin in serum, C-reactive protein, alpha-2 macroglobulin, alpha-1 glycoprotein, alpha-2 haptoglobulin – laser-nephelometric

described here by way of example. The incubation of the cytostatic agents with the xenogenic peptide preparations was carried out in the manner described in the experimental part. The laboratory values found with these patients are set out in table II.

Case Report

(1) B.-E. K., female, born 11-2-1943

Malignant thymoma, stage III. Left exploratory thoracotomy on 10-1-1983. This revealed that the pericardium, left root of the lung, mediastinum and aorta were almost completely covered and infiltrated by yellowish-white and fairly hard tumor masses.

Histology: Thymoma, stage III, which, from the proliferation tendency and spread, is behaving in a definitely malignant manner. This is particularly apparent in the tissue examined where soft tissue is infiltrated.

Figure 6 shows the mediastinal tumor of this patient with involvement of the left root of the lung, pericardium, mediastinum and aorta. This picture was taken on 2-2-1983. In figure 7 (of 30-5-1983) and figure 8 (of 5-9-1983) a regression of the tumor masses in the region of the root of the lung, aorta, mediastinum and pericardium is apparent, while in figure 3 the

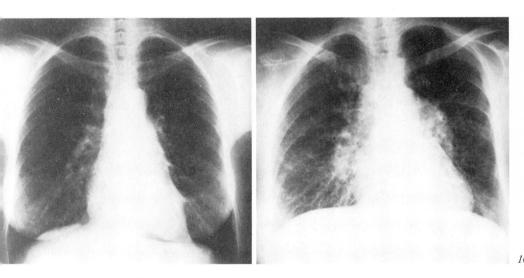

Fig. 8. B.-E. K. Follow-up of the malignant thymoma under cytobiological-cytostatic combination therapy with NeyTumorin®, mitomycin, and adriblastin. Complete regression of malignant thymoma.

Fig. 10. V. T. Advanced dedifferentiated, follicular thyroid carcinoma with lymphangiosis carcinomatosa of the lung and lymph-node metastases in the mediastinum and on both sides in the hilum.

tumor can no longer be identified. For the further assessment of the tumor regression, a CT-scan was carried out (7-10-1983) with CM bolus (fig. 9). There is a complete regression of the tumor masses at the aorta and pulmonary artery, in the mediastinum and at the pericardium. No indication of a recurrence of the tumor up to the present time.

(2) V. T., female, born 11-2-1933

A strumectomy of the right lobe of the thyroid with lymph-node extirpation was carried out with this patient in March 1983 on account of an histologically confirmed, advanced, dedifferentiated, follicular thyroid carcinoma with regional lymph-node metastases. In December 1982, hospitalization on account of a generalized formation of bone metastases, a lymphangiosis carcinomatosa of the lung and lymph-node metastases in the mediastinum and in the hilus on both sides (see fig. 10 of 28-12-1982 and fig. 11 of 9-2-1983). The patient was put on the cytobiological-cytostatic combination therapy as of 17-3-1983. The picture of 3-9-1983 (fig. 12) shows a complete regression of the lymphangiosis carcinomatosa of the lung and a regression of the lymph-nodes in the mediastinum on both sides of the hilum, apart from a lymph-node at the level of the vena acygos, which has likewise become smaller in comparison with the pictures of 28-12-1982 and 9-2-1983.

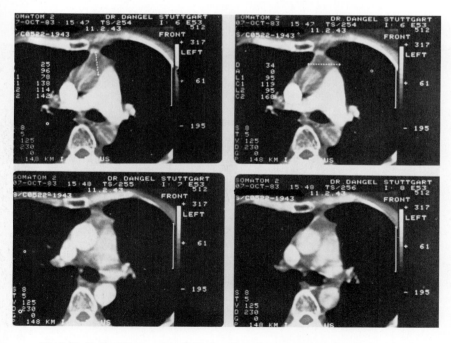

Fig. 9. B.-E. K. CT-scan with CM bolus. Complete regression of malignant thymoma, the yellowish-white and fairly solid tumor masses, which had completely covered and infiltrated the pericardium, the left root of the lung, the mediastinum and the aorta.

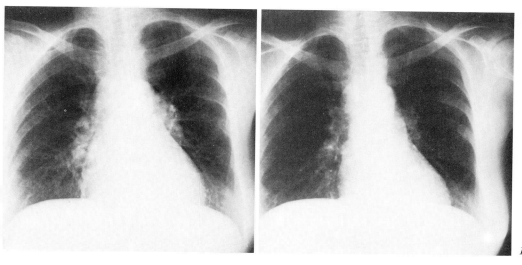

Fig. 11. V. T. Follow-up of the advanced, dedifferentiated, follicular thyroid carcinoma under cytobiological-cytostatic combination therapy with NeyTumorin®, mitomycin and adriblastin.

Fig. 12 shows a complete regression of the lymphangiosis carcinomatosa of the lung and regression of the lymph-nodes in the mediastinum and on both sides in the hilum apart from one lymph-node at the level of the vena acygos which, in comparison with the previous pictures, has likewise become smaller.

In the whole-body bone scan (fig. 13), it is apparent that no increase has taken place in the bone metastases in the period from March 1983 to February 1984. Up to the present time, there is no sign of a progression of the disease.

(3) K. G., male, born 9-7-1944

With this patient, a lymphogranulomatosis of the nodular-sclerosing type was known since February 1983. Clinical stage II with B-symptoms (staging laparotomy). After the establishment of the diagnosis, hospitalization and commencement of a polychemotherapy according to De Vita, consisting of cyclophosamide, vincristine, procarbacin, and prednisone. After an initially good response to the polychemotherapy, B-symptoms occurred once more after the 2nd cycle, as a sign of the progression and there was an increase in the size of the left supraclavicular, left axillary and mediastinal lymph-nodes (fig. 14 of 8-2-1983). Since the patient refused further cytostatic therapy, as usually at that time, and a proposed radiation treatment, a cytobiological-cytostatic combination therapy was begun with him on an outpatient basis. Under this therapy, there was a complete remission of the Hodgkin lymph-nodes in the mediastinum, in the left supraclavicular fossa and in the left axillary fossa (fig. 15 of 2-9-1983). A remission of the disease continues up to the present time.

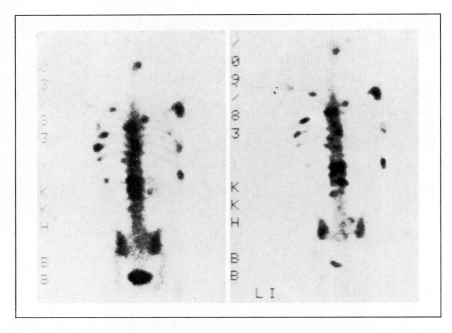

Fig. 13. V. T. Whole-body bone scans of 17-3-1983 and 3-9-1983. Advanced, dedifferentiated, follicular thyroid carcinoma. No increase in bone metastases.

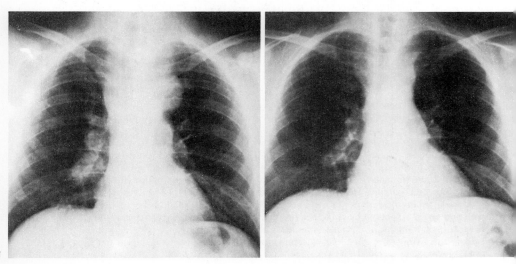

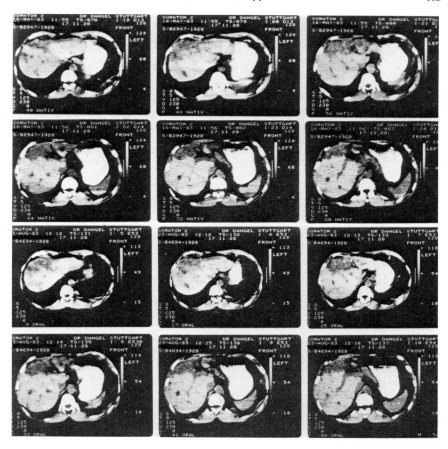

Fig. 16. J. W. Mamma carcinoma. CT-scan of the organs in the epigastric organs: diffuse permeation of both right and left liver lobules with multiple metastases (1). Under the cytobiological-cytostatic combination therapy, a progressive formation of liver metastases cannot be demonstrated (2).

Fig. 14. K. G. Lymphogranulomatosis of the nodular sclerosing type in the mediastinum affecting the hilar and mediastinal lymph-nodes.

Fig. 15. K. E. Lymphogranulomatosis of the nodular sclerosing type. After cytobiological-cytostatic combination therapy with NeyTumorin®, adriblastin, and mitomycin, almost complete remission of the Hodgkin lymph-nodes in the mediastinum.

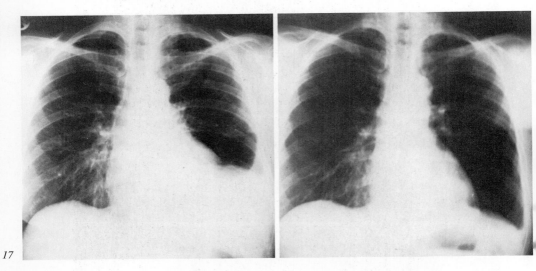

Fig. 17. J. W. Mamma carcinoma. Left pleural effusion with pleuritis carcinomatosa.
Fig. 18. J. W. Under the cytobiological-cytostatic combination therapy with NeyTumorin®, mitomycin, and adriblastin, complete regression of pleural effusion.

(4) J. W., female, born 17-11-1928

With this patient, a left ablatio mammae was carried out in 1978 on account of a mamma carcinoma with lymph-node metastases. A pleuritis carcinomatosa of the left thorax was known since April 1982. On 9-9-1982, a CT-scan of the liver revealed a hepatomegaly with diffuse permeation of both right and left liver lobules with multiple metastases. A CT-scan carried out on 16-12-1982 showed small, large and in some cases, roundish liver metastases and a left pleural effusion with pleuritis carcinomatosa. CT-scans carried out on 16-5-1983 (fig. 16) and 23-8-1983 (fig. 16) do not show a progression in the formation of liver metastases. A hepatomegaly is no longer detectable. The left pleural effusion (fig. 17 of 14-10-1982) shows a complete regression (fig. 18 of 8-10-1983).

Since the start of the therapy, a progression in the disease can no longer be demonstrated.

Fig. 19. M. S. Solidly growing, dedifferentiated, relatively small-cell, scirrhoid carcinoma of the left breast. CT-scan of epigastrium: several roundish liver metastases of varying size in the ventral region of the left liver lobule and ventrolateral and dorsolateral section of the right liver lobule. Enlarged liver.

Fig. 20. M. S. Solidly growing, dedifferentiated, relatively small-cell scirrhoid carcinoma of the left breast. After cytobiological-cytostatic combination therapy with NeyTumorin®, mitomycin, and adriblastin, complete regression of liver metastases.

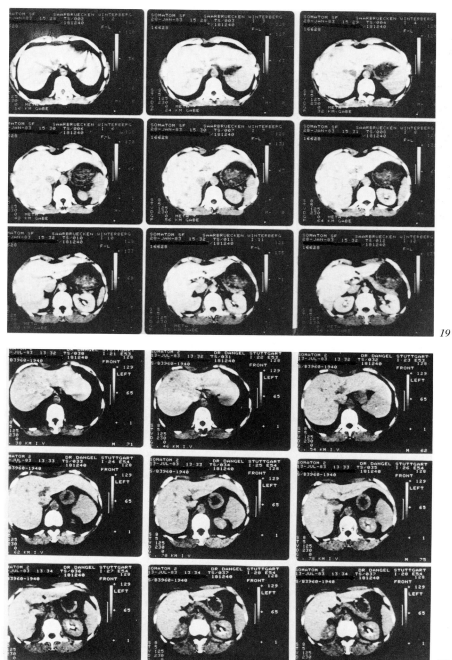

(5) M. S., female, born 18-12-1940

A left mamma ablatio with removal of the axillary lymph-nodes on account of a mamma carcinoma with lymph-node metastases was carried out with this patient in December 1982. The histology revealed a solidly growing, dedifferentiated, relatively small-cell, scirrhoid carcinoma. After establishing the diagnosis, a polychemotherapy, according to Salmon and Jones, was started. At the same time, the patient, in whom the estrogen-receptor determination in the tumor tissue was positive, was given the anti-estrogen tamoxifen, in a daily dosage of 2 × 2 tablets. Under this therapy, a formation of liver metastases occurred which was confirmed by a CT-scan (fig. 19 of 28-1-1983) and by an exploratory laparotomy with tissue biopsy. The histological examination revealed a liver metastasis of a scirrhoid carcinoma. The CT-scans of 28-1-1983 show several roundish liver infiltrations of varying size in the ventral region of the left liver lobule and ventro-lateral and dorso-lateral section of the right liver lobule. As a whole, the liver is enlarged. A cytobiological-cytostatic combination therapy was then carried out in accordance with the plan indicated above. In a control examination on 13-7-1983 (fig. 20), it was no longer possible to visualize the liver metastases. At present, there is no sign of a progression of the disease.

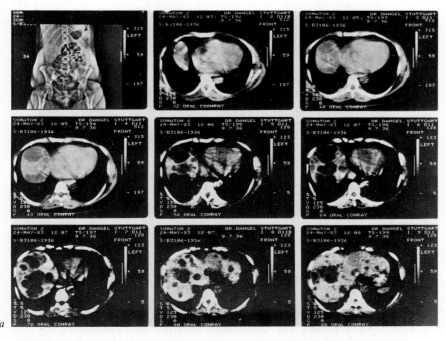

Fig. 21, a–c. A. S. Low-differentiation squamous cell carcinoma of collum uteri with no or scarcely any hornification. CT-scan of epigastrium. Under cytobiological-cytostatic combination therapy with NeyTumorin®, mitomycin, and adriblastin, clear regression in the formation of liver metastases, especially of the large metastases.

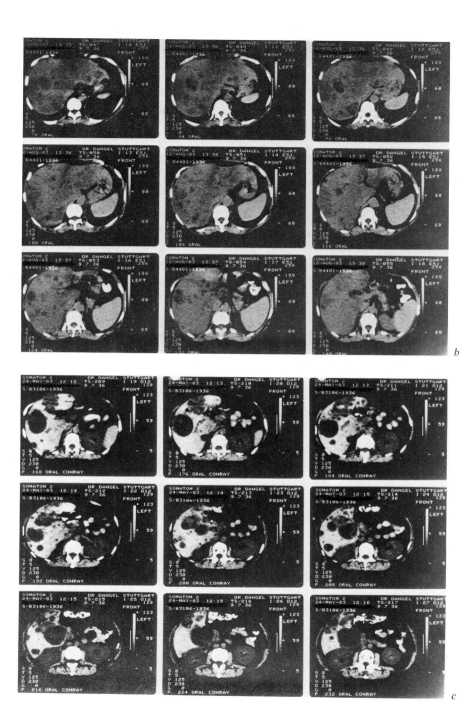

(6) A. S., female, born 9-7-1936

The establishment of the diagnosis with this patient revealed a collum carcinoma with multiple metastases. The tumor was histologically classified as a squamous cell carcinoma with low differentiation and no, or scarcely any, hornification. Figure 21, a–c, shows CT-scans of a diffuse formation of liver metastases, the size of kidneys. These pictures were obtained on 24-5-1983. Figure 22, a–c, shows a CT-scan of the same liver taken three months later after cytobiological-cytostatic combination therapy. There is a clear regression in the formation of metastases, especially with regard to the large metastases. Sudden exitus letalis in January 1984, with renewed progression of the liver metastases.

These case reports show that it is definitely possible with cytobiological-cytostatic combination therapy to achieve a stabilization of metastases formation over a period of months thus, prolonging life in a more tolerable manner. In addition to the antitumoral efficacy on this therapeutic concept, special mention should be made of the good subjective general

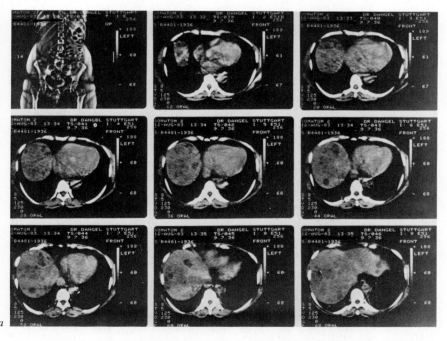

Fig. 22, a–c. After the cytobiological-cytostatic combination therapy; clear regression in the formation of metastases, above all of the big metastases. Computer tomograms of the same liver 3 months later.

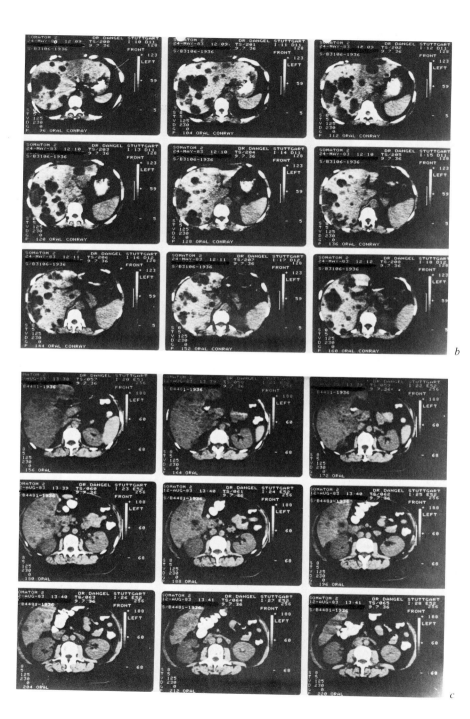

condition of the patients, which most certainly makes it possible to carry out a high-dosage cytostatic treatment on an outpatient basis.

The results documented in this paper are an encouragement for us to continue an energetic application of this principle in the therapy of solid tumors.

Summary

These case reports show that it is definitely possible with the cytobiological-cytostatic combination therapy, to achieve a stabilization in the formation of metastases over a period of months and thus, prolonging life in a more tolerable manner. In addition to the antitumoral efficacy on this therapeutic concept, special mention should be made of the good subjective general conditions of the patients which, therefore, makes it possible for a high-dosage cytostatic treatment to be carried out on an outpatient basis.

The results documented in this paper is encouragement for us to continue an energetic application of this principle in the therapy of solid tumors.

Zusammenfassung

Diese Fallbeispiele zeigen, daß es sehr wohl möglich ist, mit der zytobiologisch-zytostatischen Kombinationstherapie über Monate eine Stabilisierung der Metastasierung zu erreichen und damit das Leben lebenswerter zu verlängern. Besonders hervorzuheben ist neben der antitumoralen Wirksamkeit dieses Therapiekonzeptes das gute subjektive Allgemeinbefinden der Patienten, das auch eine Behandlung mit hochdosierten Zytostatika ambulant durchaus ermöglicht.

Die in der vorliegenden Arbeit dokumentierten Ergebnisse sind für uns Ansporn, dieses Prinzip in der Therapie solider Tumoren weiter intensiv zu verfolgen.

References

1 Huhn, D.: Internistische Tumortherapie. Med. Klin. Prax. 78: 16–25 (1983).
2 Liu, K.; Mittelmann, H.; Sproal, E.; Elia, E. G.: Renal toxicity in man treated with Mitomycin C. Cancer 28: 1314–1320 (1971).
3 Ratanatharathorn, V.; Baker, L.H.; Cadnapaphornchai, P.; Rosenberg, B. F.; Vaitkevecius, V. K.: Clinical and pathologic study of mitomycin C nephrotoxicity; in Carter, Crooke, Adler (eds.), Mitomycin C, pp. 219–229 (Academic Press, New York, San Francisco, London 1979).
4 Theurer, K.; Paffenholz, V.: Einfluß von makromolekularen Organsubstanzen auf menschliche Zellen in vitro. I. Diploide Kulturen. Kassenarzt 27: 5218–5226 (1978); II. Tumorzellkulturen. Kassenarzt 19: 1876–1887 (1979).
5 Letnansky, K.: Stoffwechselregulatoren der Plazenta und ihre Wirkung in Normal- und Tumorzellen. Exp. Pathol. 8: 205–212 (1973).
6 Letnansky, K.: Tumorspezifische Faktoren der Plazenta und Zellproliferation. Exp. Pathol. 9: 354–360 (1974).

7   Letnansky, K.: Inhibition of thymidine incorporation into the DNA of normal and neoplastic cells by a factor from bovine maternal placenta: interaction of the inhibitor with cell membranes. Biosci. Rep. *2:* 39–45 (1982).
8   Letnansky, K.: Entdeckung zellulärer Rezeptoren für antitumorale plazentare Faktoren in NeyTumorin. Therapiewoche *33:* 59–61 (1983).
9   Ketelsen, U.-B.: Pilotstudie zum Einfluß eines biologischen "Response Modifiers" (NeyTumorin) auf die Plasmamembran menschlicher Tumorzellen (Wish) in vitro im Vergleich zu einem Chemozytostatikum (6-Mercaptopurin). Therapiewoche *33:* 62–70 (1983).
10 Munder, P. G.: Experimentelle Untersuchungen über den antitumoralen Wirkungsmechanismus von NeyTumorin. Therapiewoche *33:* 71–73 (1983).
11 Munder, P. G.; Stiefel, Th.; Widmann, K. H.; Theurer, K.: Antitumorale Wirkung xenogener Substanzen in vivo und in vitro. Onkologie *5:* 2–7 (1982).
12 Theurer K. E.: Multifaktorielle Krebstherapie mit hochmolekularen Organextrakten und tumortropen Antikörperfragmenten. Therapiewoche *33:* 17–22 (1983).
13 Kisseler, B.; Stiefel, Th.: Eine Pilotstudie zum Verlauf der Tumormarker CEA, TPA und Ferritin bei 29 markerpositiven Patienten mit Mamma-Ca unter der Therapie mit NeyTumorin-Sol in Verbindung mit Chemotherapie. Therapiewoche *33:* 4993–5006 (1983).
14 Lindenmann, M.: Die Stellung der makromolekularen Organotherapie in der Onkologie. Kassenarzt *20:* 10 (1980).
15 Reuter, H.-J.: Die multifaktorielle immunologische Krebstherapie in der Urologie. Helv. Chirurg. Acta *43:* 279–283 (1976) und Therapiewoche *33:* 99–100 (1983).
16 Porcher, H.: Bericht über die Tagung der Gesellschaft zur Erforschung der Makromolekularen Organo- und Immunotherapie (GEMOI) (Abstract). Erfahrungsheilkunde *12:* 876–890 (1983).
17 Gurr, E. et al.: Synthetic Dyes and Biological Problems, p. 319 (Academic Press, London 1971).
18 Phillips, H. J.: Dye Exulsion Tests for all Viability; in Kruse et al. (eds.), Tissue Culture, pp. 407–408 (Academic Press, New York 1973).
19 Waymouth, C.: To Disaggregate or not to Disaggregate, Injury and Cell Disaggregation, Transient or Permanent? In-Vitro *10:* 97–111 (1974).
20 Phillips, H. J.: Dissociation of Single Cells from Lung or Kidney with Elastase. In-Vitro *8:* 101–105 (1972).

PD. Dr. med. B. Kisseler, Radiolog. Zentralabtlg. des Kreiskrankenhauses Böblingen, Bunsenstr. 120, D-7030 Böblingen (FRG)

Pilot Experience with NeyTumorin®-Sol for the Treatment of Generalized Metastasizing Carcinomas of the Mamma

O. F. Lange

Robert Janker Clinic, Bonn, FRG

Introduction

Despite all the therapeutical progress in carcinoma of the mamma, hematogenous metastatic-spread finally occurs in a large number of cases. Tumor regressions of varying extent and duration can be achieved, even in this stage, by established tumor treatments, especially by the combined application of radiotherapy and cytostatic chemotherapy. In the short or long term, however, a refractory tumor progression occurs.

This group of patients suffering from carcinoma of the mamma, for whom no further treatment is possible, represents a major medical and psychosocial problem for every oncological specialist. The efforts made so far to achieve remissions of an improvement in the quality of life, have not led to convincing results – up till now.

Selective, Antiproliferative, and Resistance-Stimulative

In this situation, cytoplasmic therapy with organ extracts may possibly indicate a new approach. In animal experimentation and human cell-cultures it was found that NeyTumorin® has a tumorostatic or tumoricidal effect, and in this respect thus resembles the cytostatic chemotherapeutic agents, but without the side-effects usually found with this group of medicaments.

An explanation for this is the suspected mechanism of action of the organ lysates:
(1) The antiproliferative effect acting selectively on the tumor cell;
(2) the stimulation of certain cell-subserved defense mechanisms.

Purpose of Study

Initially, in a pilot study, the general roborant effect and the antitumoral action of the preparation NeyTumorin® were to be studied in individual cases. Up to the present, 12 patients took part in the study. The following inclusion-criteria had been found:
(1) Histologically confirmed carcinoma of the mamma with hematogenous metastatic spread;
(2) the stimulation of certain cell-subserved defense mechanisms (radiotherapy, chemotherapy, and/or hormonal therapy).

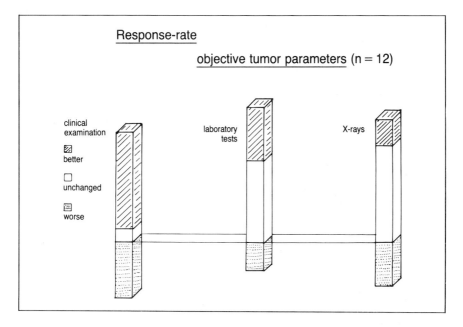

Fig. 1. Localization of metastases. Study of the palliative efficacy of NeyTumorin® in the general metastatic spread of carcinoma of the mamma after operation, chemotherapy, and radiation treatment.

Case Material

Twelve female patients (age 34 to 65 years) in the preterminal stage of carcinoma of the mamma had been treated up to this time with the NeyTumorin®-regime. Bone metastases could be demonstrated in 11 cases, liver metastases in 8 cases, pulmonary metastases in 6 cases, lymph-node metastases in 6 cases, and skin metastases in 3 cases (fig. 1).
All patients had previously received radiation, cytostatic, and hormonal treatment.

Therapeutic Regime

Day 1 morning and evening 1 ampoule NeyTumorin® each, dilution strength I;
day 2 morning and evening 1 ampoule NeyTumorin® each, dilution strength II;
day 3 morning 1 ampoule NeyTumorin®, dilution strength III, i.v. in each case;
days 4 to 8 1 ampoule NeyTumorin®-Sol i.v. each day;
days 9 to 20 1 ampoule NeyTumorin®-Sol i.m. each day.

In addition to this therapy, a symptomatic treatment with analgesics, antiemetics, and sedatives was carried out. In two cases, an additional radiation therapy was necessary on account of the danger of a pathologic fracture in the region of osteolytic bone metastases.

Assessment Criteria

The following subjective criteria were selected as assessment parameters: general condition, appetite, pain, and nausea. In addition, the following objective parameters were documented: measurable size of tumor or metastasis, weight, laboratory values, and X-ray examinations.

Results

General Condition

Eight of the 12 patients treated felt generally better and stronger, already within a few days after the start of the therapy. In 3 cases, there was no change in the general condition, even on completion of the 20-day cycle treatment. In one patient, there was a deterioration in the general state of health under the therapy.

Weight

Five patients shamed a desired increase in weight occurring during the treatment. In another 5 women, the progressive loss of weight was halted. In two cases, there was an increase in the extent of the cachexia of malignancy (fig. 2).

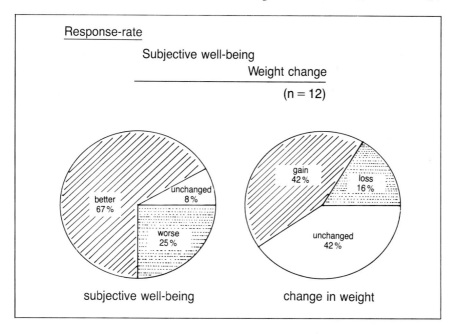

Fig. 2. Response-rate: subjective well-being, weight change (n = 12). Study of the palliative efficacy of NeyTumorin® in the general metastatic spread of carcinoma of the mamma after operation, chemotherapy, and radiation treatment.

Pain

Of 11 patients who regularly needed high-dosage analgesics, it was possible to clearly reduce, in some cases, the consumption of pain-killers by 7 of them. There was no change in the analgesic requirements of the other 4 patients (fig. 3).

Nausea

Four patients suffered from constant, severe nausea and vomiting. In one case, it was possible to discontinue, and in another significantly reduce, the antiemetics during the NeyTumorin® treatment.

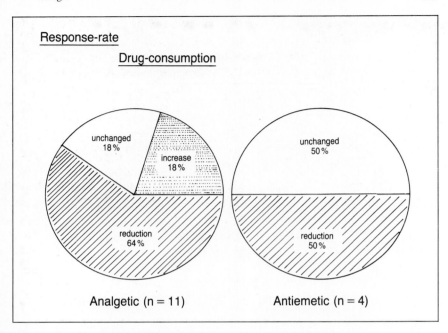

Fig. 3. Response-rate: drug-consumption. Study of the palliative efficacy of NeyTumorin® in the general metastatic spread of carcinoma of the mamma after operation, chemotherapy, and radiation treatment.

Objective Tumor Regressions

The clinical examination revealed an improvement in 7 of the 12 patients (reduction in size of skin or lymph-node metastases, reduction in the tapping pain which could be induced at skeletal areas affected by metastases, reduction in size of livers affected by metastases). With the other 5 patients, a progression had occurred in the tumor condition.

Laboratory Parameters

In the majority of the cases, there was no real change in the laboratory values during the treatment. In 4 of 12 cases, a clear tendency to normalization was found (gamma-GT, LDH, alkaline phosphatase, CEA, and BSR).

X-Ray Examinations

In one case, a complete regression of a pulmonary lymphangiosis was found and, in another case, a partial regression of pulmonary metastases (fig. 4).

Description of an Individual Case

With a female patient of 38 years suffering from bone, liver, and lymph-node metastases and an extensive pulmonary lymphangiosis carcinomatosa, the NeyTumorin® treatment was started in April 1983, after all other therapeutic possibilities had been exhausted. The patient was in the preterminal stage and was suffering from very severe dyspnoea at rest. There was a clear improvement in her state of health within a few days after the start of the therapy. Sonographic examination showed a regression in the liver metastases, while X-rays revealed an improvement in the

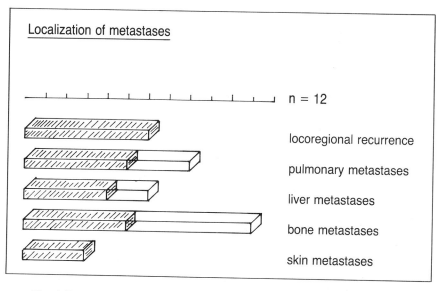

Fig. 4. Response-rate: objective tumor parameters (n = 12). Study of the palliative efficacy of NeyTumorin® in the general metastatic spread of carcinoma of the mamma after operation, chemotherapy, and radiation treatment.

lymphangiosis carcinoma. The patient was discharged with freedom from pain after three weeks.

At the wish of the patient, the NeyTumorin® was continued on an outpatient basis by the family doctor. With continuation of the treatment, the remission lasted for four months. After this, a progressive growth of the tumor took place again.

With the other patients who did not receive further treatment after the 20-day therapy cycle with NeyTumorin®, the improvement lasted for 1 to 7 weeks.

Improvement in Quality of Life

During and after the cytoplasmatic therapy in the dosage quoted above most patients confirmed an improvement of physical well-being and mental activity. This therapeutic effect is of great importance, especially in far advanced cancer patients who are not able to receive any further treatment (hormones or cytotoxic drugs). The improvement in quality of life lasted up to 7 weeks after the end of NeyTumorin® therapy.

New Developments

Because of the unspecific immunostimulating and possibly antitumoral effect of xenogenic peptides NeyTumorin® will be used in multimodal cancer therapy as a supportive treatment in combination with radiation and cytotoxic drugs. In some cases the xenogenic peptides have been able to reduce the side-effects of aggressive polychemotherapy regimes.

This will be the subject of further investigations. A prospective randomized study will be performed.

Summary

The results of the pilot study show, that in terminal cases of metastasizing carcinomas of the mamma, a therapy with NeyTumorin® alone, in the dosage quoted, leads in some cases to an improvement in mental and physical well-being and to better appetite and functioning capacity. Also, a desirable gain in weight can be achieved with cachectic patients. In many cases, it is possible to reduce the analgesics and antiemetics administered. In individual cases, an objective regression in the tumor condition could be shown by clinical, laboratory, and X-ray findings. No side-effects occurred in the patients treated, apart from an occasional rise in the heart-rate during the intravenous injection of NeyTumorin®-Sol. Special mention should be made of the improvement in the quality of life, which is often already reported within a short time of the therapy. The improvements achieved lasted for 1 to 7 weeks. It is probable that the remission times are longer when NeyTumorin® treatment is continued after the 20-day therapy cycle.

Zusammenfassung

Die Ergebnisse der Pilotstudie zeigen, daß in terminalen Fällen metastasierender Mammakarzinome eine alleinige Therapie mit NeyTumorin® in der angegebenen Dosierung in einigen Fällen zur Besserung des physischen und psychischen Allgemeinbefindens sowie zur Steigerung von Appetit und Leistungsfähigkeit führt. Eine erwünschte Gewichtszunahme ist auch bei kachektischen Patienten noch zu erreichen. Analgetika und Antiemetika lassen sich oftmals reduzieren. In Einzelfällen war eine objektivierbare Rückbildung des Tumorleidens klinisch, laborchemisch und röntgenologisch feststellbar. Bis auf gelegentlichen Anstieg der Herzfrequenz während der intravenösen Injektion von NeyTumorin®-Sol traten bei den behandelten Patientinnen keine Nebenwirkungen auf.

Besonders zu erwähnen ist die Verbesserung der Lebensqualität, die oft schon kurz nach Therapiebeginn angegeben wird. Die erreichten Besserungen hielten 1–7 Wochen an. Es ist wahrscheinlich, daß die Remissionszeiten unter Fortführung der NeyTumorin®-Behandlung über den 20tägigen Therapiezyklus hinaus länger andauern.

Dr. med. O. F. Lange, Robert-Janker-Klinik, Baumschulallee 12, D-5300 Bonn (FRG)

Macromolecular Organ Extract (NeyTumorin®) in the Treatment of Non-Small-Cell Bronchial Carcinoma and Metastatic Lung Disease
Preliminary Report

K. H. Bohnacker, F. Krause

Dept. I with Thoracic Surgery, Löwenstein Lung Clinic, Löwenstein, FRG

NeyTumorin® (NT) is reported to possess cancerostatic properties [3–5]. This was to be tested by a pilot study in patients with advanced primary non-small-cell bronchial carcinoma (NSBC) and with secondary metastatic lung disease in regard to dose, application, tolerance, and side effects.

Patients and Methods

From November 1981 until September 1982 NT was given to 10 patients with histologically proven NSBC, partially with limited disease and to patients with metastatic lung cancer of extrathoracic origin; all showing progression of their disease. Some patients were treated previously, with more or less effect. In all cases, surgery was not possible anymore or had been refused, and further polychemotherapy courses did not seem to benefit the patients and could be postponed (see table II).

NT was applied according to the instructions of the manufacturer. It was planned to administer NT for at least 6 weeks as mono- and only therapy. Before and after 6 weeks of treatment, the following parameters were to be obtained:
Subjective feeling,
acceptance of the treatment and the injections,
vigilance and performance,
weight, sedimentation-rate, WBC, IgE, CEA, roentgenology.
Administration was performed i.m. beginning with dilution no. 66 for the first 3 days followed by NT as dry substance. The time interval between doses being not longer than 3 or 4 days. After day 19 NT was given twice a week on Mondays and Thursdays, as dry substance (table I).

Table I. Dosage and timing of NeyTumorin®

Day 1	NT dil. no. 66 St. I	2/daily
Day 2	NT dil. no. 66 St. II	2/daily
Day 3	NT dil. no. 66 St. III	1/daily
Day 4	NT dry substance 15 mg	1 amp.
Day 7	NT dry substance 15 mg	2 amp.
Day 10	NT dry substance 15 mg	2 amp.
Day 13	NT dry substance 15 mg	2 amp.
Day 16	NT dry substance 15 mg	2 amp.
Day 19	NT dry substance 15 mg	2 amp.

From the 19th day on, twice weekly, for instance Mondays and Thursdays, 1 amp. NT dry substance. This could also be done on an outpatient basis.

Results

Ten patients were treated with NT. Two female patients died of their disease in the clinic during treatment. Both had lymphangiosis carcinomatosa and rapidly progressing tumor disease. Cause of death was cachexia and respiratory failure.

The remaining 8 patients and their follow-up data are summarized in table III. All patients died of their respective disease after discharge from hospital. One patient had radiotherapy after NT.

Vigilance and Subjective General Condition

Two patients reported feeling much better after the first couple of injections; all others did not report any change. One patient reported feeling worse, but this was in accordance with his failing tumor condition. One patient with lung metastases continued having a good general condition, in spite of his ever-growing lung metastases.

Radiological Changes

In all patients but one (no. 6), there was a continuous enlargement of the roentgenological intrathoracic tumor signs. The above mentioned patient was first staged with "no change" intrathoracically, but had tumor progression in the abdomen.

Table II. Synopsis I of all patients before NeyTumorin® medication

No.	1 Initials	Age Sex y/m/f	1 Tumor type	2 Stage TNM	3 Weight kg	4 Performance*	5 Sediment rate 2 h	6 Serum CEA	7 Previous treatment	8 Date of diagnosis month/year	9 NT Month post diagnosis	10 Remarks
1	Z.J.	58/m	Squ.c.	$T_3N_2M_0$	60	2	114	5	Cyt.	9/81	4	
2	K.W.	76/m	Squ.c.	$T_2N_2M_1$	55	2	110	60	–	4/82	0	
3	V.E.	41/m	Squ.c.	$pT_3N_2M_1$	56	1	34	9	Op + Cyt.	3/75	76	**
4	J.E.	77/m	Large cell	$T_1N_0M_0$	76	0	12	5.8	R. Th.	4/82	0	***
5	R.K.	62/m	Large cell	$T_3N_2M_0$	55	2	129	3.7	Cyt.	5/80	21	
6	J.E.	71/m	Alv.c.	$pT_2N_2M_0$	70	1	9	24.4	Op + Cyt.	2/81	11	
7	S.P.	69/m	Adeno cell	$T_2N_2M_0$	59	2	49	8.6	–	11/81	2	
8	S.F.	60/m	Parotid gland	no TNM	67	1	48	12.3	Op + R. Th.	9/81	5	
9	S.G.	64/f	Breast carc.	no TNM	58	4	–	–	–	–	–	Died suddenly during NT
10	R.W.	72/f	Tu of tub.ov.	no TNM	63	4	–	–	–	–	–	Died during NT treatment

* Moertel, C. G.; Cancer Chemoth. Rep. 58: 257–259 (1977)
** Denied radioth. after resection at stage $pT_2N_2M_0$, then tumor was slowly progressing
*** Denied Op. treatment, chemoth. and radioth., since of experimental character, postponed

Tumor-Associated Antigens IgE and CEA

In all patients, serum CEA levels did not rise in accordance to the severeness of their tumor progression. During NT administration serum CEA levels remained rather stable or showed a falling tendency (see figs. 1, 2) IgE levels remained rather stable as did all other blood parameters.

White and Differential Blood Count

There was no remarkable change in the relation leucocytes/lymphocytes. In one case, eosinophilic count fell under NT treatment. Numbers of WBC, however, remained stable as did total blood proteins and the respective body weights.

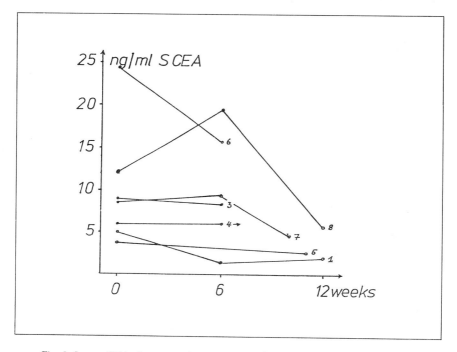

Fig. 1. Serum CEA changes under NeyTumorin® therapy in recordings before starting and after 6 weeks, or before patients' final failure. Missing patient no. 2 had values over 60 ng/ml and is not recorded.

Table III. Synopsis II follow-up of all patients under and after NeyTumorin® medication

No.	Initials	1 Weight loss/kg	2 Sediment rate	3 Serum CEA	4 X-ray follow-up	5 Side effects	6 Patients' acceptance* − ± +	7 Physicians Opinion** Performance	8 Survival months after diagnosis	9 Survival months after NT	10 Remarks
1	Z.I.	2	↑	→	progress	−	±	2 3	10	6	
2	K.W.	2	→	−	progress	−	−	2 3	3	3	
3	V.E.	1	↑	→	progress	−	±	1 1	83	7	1975 at diagnosis Patient had $T_2N_2M_0$
4	J.E.	1	↑	→	progress	−	+	0 0	13	13	
5	R.K.	−	↑	→	progress	−	+	2 3	26	5	
6	J.E.	−	↑	→	no change	−	+	1 1	14	3	
7	S.P.	1	↑	→	progress	−	+	2 3	6	4	
8	S.F.	2	→	→	progress	−	±	1 1	19	12	***

* Patients' opinion about the effect of NT injections
** Physicians' opinion about general condition and performance using Moertel scale
*** Follow-up see figure 2

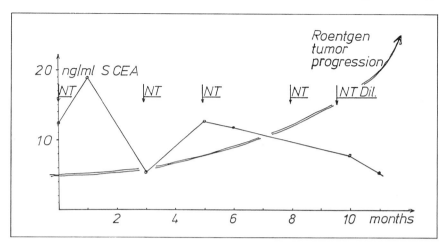

Fig. 2. Follow-up of CEA levels of patient no. 8 with carcinoma of parotid gland and lung metastases. CEA recordings under NT therapy fell to normal values, inspite of tumor progression when leaving our control.

Side Effects

There were no side effects seen, whatsoever.

Length of Survival

Since there were different pre-treatment courses, different TNM-stages and no so-called hard data, considerations in this regard are more or less speculative. However, there seems to be a tendency of amelioration of quality of survival or at least, some kind of stabilization of the general condition under NT treatment.

Discussion

Because all patients but one (no. 4), suffered from rather far-advanced cancer, some of them already being multimodally treated, it seemed to be quite optimistic to expect effects on the course of the ever-progressing disease or on survival time. But nevertheless, in this respect, the subjective feeling of good general condition, inspite of tumor progression, is remarkable.

In regard to dose and application, there were no problems which could be associated to the xenogenous origin of the organ extract. Whether the upper border of the tolerable amount of the substance had been reached, or whether a dose/effect relation exists, could not be elucidated. Allergic reactions were not observed. The tolerance against a 20-fold dose of the substance has, in the meantime, been published [1].

The duration of the medication was supposed to be long enough to demonstrate some effect on the tumor disease. Here it is interesting that until now, the fact that CEA-levels remained fairly equal under treatment with NT, inspite of tumor progression (mentioned also from other observers), has been unexplained. Whether this is a direct influence of the substance on the tumor cells and/or their antigen production caused by some biologic reactions, remains to be investigated.

If immunology is supposed to play any role in the spectrum of effects of NT, there should be some change of the relation of neutro- to lymphocytic-blood-count which, however, was not observed. Therefore, any statement in this respect is also very speculative. There are very many questions still to be solved.

Summary

NeyTumorin®, an xenogenous organ extract, has been given to a small number (10) of non-small-cell bronchial carcinoma patients and some with metastatic lung disease of extrathoracic origin. All patients but one had been pretreated with multimodal therapy. Aim of this small pilot study was the questioning of antineoplastic effects of the substance as well as for dosage, tolerance, administration, and side effects. The usual parameters for a follow-up of cancer patients were obtained. There was no effect seen on intrathoracic tumor growth, except in one case. Slightly lowered levels of CEA with a falling tendency, inspite of roentgenological tumor progression were, however, observed but could not be explained. There were no side effects.

Zusammenfassung

Das xenogene Präparat NeyTumorin® wurde bei einer kleinen Zahl von nicht-kleinzelligen Bronchialkarzinompatienten und Patienten mit Lungenmetastasen extrathorakalen Ursprungs im Rahmen einer klinischen Pilotstudie eingesetzt. Außer einem Patienten wurden sämtliche Patienten dieser Studie mit einer multifaktoriellen klassischen Therapie behandelt.

Ziel dieser Pilotstudie war neben dem Nachweis einer antineoplastischen Wirkung von Ney-Tumorin® die Untersuchung dieses Präparates bezüglich Dosierung, Verträglichkeit, Art der Anwendung und Nebenwirkungen. Im Rahmen der Studie wurden die gebräuchlichen Parameter eines Follow-up bei Tumorpatienten erhoben. Bezüglich des intrathorakalen Tumorwachstums konnte nur bei einem Patienten ein therapeutischer Effekt nachgewiesen werden. Geringfügig erniedrigte CEA-Werte mit fallender Tendenz wurden trotz röntgenologischer Tumorprogression beobachtet. Nebenwirkungen konnten in der gewählten Dosierung bei keinem der Patienten beobachtet werden.

References

1. Kisseler, B.; Stiefel, Th.: Pilotstudie zum Verlauf der Tumormarker CEA, TPA und Ferritin. Therapiewoche *33:* 4993–5006 (1983).
2. Letnansky, K.: Inhibition of thymidin incorporation into the DNA of normal and neoplastic cells by a factor of bovine maternal placenta. Interaction of the inhibitor with cell membranes. Biosci. Rep. *2:* 39–45 (1982).
3. Munder, P. G.; Stiefel, Th.; Widmann, K. H.; Theurer, K.: Antitumorale Wirkung xenogener Substanzen in-vivo und in-vitro. Onkologie *5:* 68–75 (1982).
4. Munder, P. G.: Experimentelle Untersuchungen über den antitumoralen Wirkungsmechanismus von NeyTumorin®. Therapiewoche *33:* 71–73 (1983).
5. Pfaffenholz, V.; Theurer, K.: Einfluß von makromolekularen Organsubstanzen auf menschliche Zellen in-vitro. Der Kassenarzt *19:* 1876–1887 (1979).

Dr. med. K. H. Bohnacker, Oberarzt der Klinik Löwenstein, D-7101 Löwenstein (FRG)

Active Immunotherapy in Metastasizing Hypernephroma

K. F. Klippel

Urological Clinic, General Hospital, Celle, FRG

Hypernephroma metastases are the most common radio-resistant tumors; 90% are not effected by cytostatic drugs. It was impossible to gain a prolonged life-time in patients tested with gestagen-preparations. Once metastasizing has set in people are generally expected to live for another 5–7 months [4]. None of the patients treated by cytostatic drugs, radiotherapy and surgery lived past the 5-year limit.

Tykkae [13] was the first to report on the total remission of metastasizing hypernephroma using immunotherapy. In 1978, the same group reported on a randomized study, effecting metastasizing hypernephroma. After excision of the tumor, patients were injected with autologous tumor cells, together with BCG or candida antigens, which are unspecific immunostimulators. 23.6% of these patients lived past the 5-year limit, while only 4.3% of the conventionally treated patients stayed alive. This impressing result started a pilot study in 1976; results of which are detailed here.

Patients and Methods

We are talking about 7 patients (table I); 4 female and 3 male in good general condition and an average age of 44.8 years, whereby surgical measures would have been of no help because of localization of metastases. The side-effects of the vaccinations were well-tolerated in all patients except patient no. 1, E. S. Out of 4 living patients, 2 are able to do a full-time job. The patients were told about the stage of their tumor and that the treatment given was still in its experimental stage.

Table I. Results of pilot study

Pat./age		Sex	OP	Metastases	Start immunother. postop.	Cells	C.R.	P.R.	Progress	Follow-up	Status
(1)	E.S. 39	f	3/77	lung	2 weeks	auto.	∅	∅	yes	9 months	dead
(4)	O.W. 62	m	11/76	bone	27 months	allo.	∅	∅	yes	34 months	dead
(5)	W.J. 44	m	Emb. 4/78	bone lung	13 months	allo.	∅	∅	yes	22 months	dead
(2)	K.R. 50	f	12/76	lung, CNS, bone	12 months	allo.	∅	yes	∅	91 months	at risk (60%)
(3)	E.J. 37	m	11/78	lung	2 weeks	auto.	yes	∅	∅	67 months	at risk (100%)
(6)	G.Sch. 44	f	3/79	lung, liver	5 weeks	auto.	yes	∅	∅	64 months	at risk (100%)
(7)	M.F. 48	f	6/75	lung, bone, kidney	5 years	allo.	∅	yes	yes	108 months	at risk (80%)

C.R.: complete remission;
P.R.: partial remission;
percent: Karnofsky index;
Auto.: autologous cells;
Allo.: allogenic cells

Cell Collection

The tumor cells were removed under operation, put into APMI 1640 medium (12.5% horse serum, 2.5% foetal calf serum, penicillin, streptomycin, hepes-buffer solution and glutamine) kept at 25° C, in Falcon-flasks. They were used within 60 min of collection. (80-Mash, trypsinization, eosine-exclusion test, preparation of tumor-cell line, shock-freezing for storage of cell rests in tumor bank at $-172°$ C).

Before injection of the cells they were defrosted, counted, a vitality test done and irridiated with 10 000 R/Cobalt rays.

Immunization

5×10^6 to 10×10^6 autologous or allogenic tumor cells were injected together with the immunostimulator, Keyhole Limpet Hemocyanin (KLH) (Calbiochem 374805, A grade), or with candida antigens (Candidin 1:100). Patients who were treated with immunostimulator KLH (0.5 mg), were treated or presensitized with KLH 4 weeks previously.

Skin Reactions

6–12 h after treatment a localized erythema was noted at injection site (left lower arm) together with slight puritic effects and subfebrile temperatures. After the injections, the axillary lymph nodes in nearly all patients were palpable as big stringy objects. All these acute reactions disappeared 5–6 days after the injections. The local induration, partly with hyperpigmentation went after 4–6 weeks.

Choice of Patients

Patients were picked at random. Written consent for therapy was given by all patients. The following parameters, immunologically were tested:
(1) Electrophoresis
(2) Immunoglobulin
(3) T-cells (Rosette-test)
(4) B-cells (immunofluorescent test with anti-light chain IG; SS
(5) DNCB-test
(6) Lymphocyte stimulations
  – Phythemagglutinin (PHA)
  – Pokeweedmitogen (PWM)
  – Concanavalin A (Con A)
(7) Candida-antibody titer
  – Immunofluorescence
  – Hemagglutination titer (Hoffman-La Roche).

All patients were seen on a regular basis in the tumor clinic. During those visits routine tests, such as full blood-count (Fbc), electrolytes, phosphatases, creatinine and protein, CEA and urine analysis were taken out. Chest X-rays were made monthly or 3-monthly, urograms were made half-yearly, the whole-body bone scan yearly and, if necessary, further check-ups like computer tomogram or ultrasound were made.

Case Histories

(1) E. S., female, 39 years

After pre-operative radiation with 4000 rad, stage of tumor $P_3N_0M_2$ a total tumor nephrectomy was carried out in March 1977 (left kidney). The patient was given daily injections into the right thigh, of 5×10^6 tumor cells plus 5 mg KLH, for 14 days following the operation. Monthly routine chest X-rays showed stabilization of metastases for 4 months. Because the patient had a very severe local reaction (erythema 4×8 cm) and painful induration, she refused further injections. Polychemotherapy was started in July 1977. Following this, metastases in the lung spread rapidly and the patient died in December 1977.

(2) K. R., female, 50 years

After pre-operative radiation with 4000 rad, stage of tumor $P_3N_0M_2$ a total tumor nephrectomy L was carried out in December 1976. Two lung metastases were first seen in an X-ray taken in October 1977. First injections of 5 mg KLH (intradermal) were given in December 1977. This was followed by injections of 4×10^6 allogenic tumor cells plus 5 mg KLH in January 1978. An X-ray taken in February 1978 showed shrinking of metastases, while in March 1978, metastases were not visible and again not visible in April.

Tomography of the lungs done in December 1978 showed no sign of metastases. A booster of 1 mg KLH was given intradermally, followed by a strong skin reaction. She had an apoplectic attack in the late part of October which left her with hemiparesis of the left side. The computer tomogram showed a single brain-metastasis with no localization in EMG. Furthermore, it showed enlargement of right single kidney, the whole-body bone scan showed negative, the mediastinal tomogram showed enlarged lymph nodes, which could possibly be metastases. There is now no known localization of metastases, and 6 years after immunization, the patient is still alive with a known brain metastasis, an enlarged kidney, and possible lymph-node metastases.

(3) E. J., male, 37 years

After pre-operative radiation with 3600 rad, stage of tumor $P_1N_2M_2$ (lung metastases) a total tumor nephrectomy R was carried out in November 1978. After immunization with 5 mg KLH (intradermal), an injection of 2×10^6 syngenic cells followed in December 1978. January 1979, no further metastases showing on chest X-ray. In February 1979, however, after a second injection of 3.5×10^6 plus 1 mg KLH, lung metastases were showing in their original location and in greater numbers. Status of multiple pulmonary metastasizing was reached in April 1979.

Regression of all metastases in December 1980, 2 years after starting the immunotherapy. The patient now is in very good general condition and able to do a full-time job. In June 1983 a solitary lesion in the lung appeared, which was operated and used for new vaccination.

(4) O. W., male, 62 years

After orthopedically treating a spontaneous fracture L humerus with known osteolysis of the scapula, an operation was carried out only to find a hypernephroma R.

After pre-operative radiation with 4000 rad a total tumor nephrectomy R was carried out in November 1976 (stage of tumor $P_1V_1N_0M_{1D}$).

Because of osteolysis in the left shoulder-joint, an exarticulation had to be performed, as there were still more osteolyses in the scapula and several ribs. Following this, the patient was presensitized with 5 mg KLH intradermally. He was injected with 5×10^6 allogenic hypernephroma cells plus KLH in February 1979. Despite this, metastases kept progressing so that the patient died, 6 months after starting immunotherapy and 3 years after palliative tumor nephrectomy.

(5) P. J., male, 44 years

After extirpation of a single metastasis located at the front of the skull-cap, the following EEG and tomography of the skull showed no irregularity.

Because hypernephroma R was seen after angiography and multiple metastases were known to be in lung and bones, embolization of the right kidney-artery with histoacryl was carried out. The number of metastases in the lung doubled between April 1978 and August 1978. Number of metastases stayed stable after immunization with 5 mg KLH given in December 1978. January 1979 saw rapid growth of lung metastases and relapse in front skull-cap.

Computer tomogram showed atrophy of the brain with no sign of metastases. The patient became tachycardiac and dispnoeic, as a result of enormous growth of metastases. Injection of 10×10^6 allogenic-radiated hypernephroma cells plus KLH was given intradermally on 1st February 1979. Added treatment with testosteron and digitalis saw stabilization of metastases in December 1979. The patient died February 1980 following a cancerous cachexia.

(6) G. Sch., female, 44 years

A total tumor nephrectomy was carried out 20th March 1979 (fig. 1, left kidney). During the operation, a single metastasis was seen in the liver while there were multiple metastases known to be in the lungs. Histopathology showed cancer stage $P_3G_3V_1M_{1d}$ ("bad cancer"). Metastases in the lungs increased significantly in size and number 4 weeks after the operation. Injection of 5×10^6 autologous tumor cells plus 0.2 ml Candidin was given 5 weeks after the operation.

Slight skin reactions, Ig-titer against candida albigans 1:80, IgM 1:640. May 1979 restimulation with candida-antigene 1:10, after that candida titer IgG 1:640, IgM 1:1.280.

Computer tomogram of liver July 1979 shows liver metastasis still existing. August 1979 restimulation with 5×10^6 autologous cells plus Candidin 1:100 given intradermally. 3-monthly restimulation with autologous cells till December 1980. The running computer tomograms of the liver and 6-weekly chest X-rays show stabilizing of lung metastases till August 1980 with cystic change of known liver metastasis. Continuous regression of lung metastases after this. After numerous chest X-rays in December 1980 no metastases were seen in the lungs. The patient gained 10 kg in weight, is feeling physically fit and doing a full-time job. In February 1982 the liver lesion was punctured by the radiologist: no signs of tumor cells. The patients is still doing well.

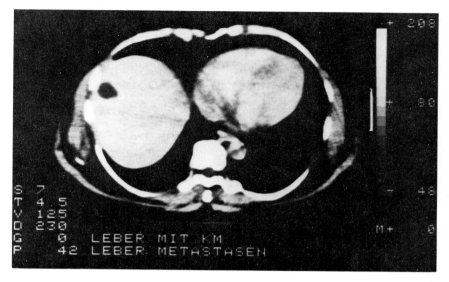

Fig. 1. Computer tomography of patient No. 6, G. Sch. She had an intraoperatively-verified liver lesion. In serial CAT-scans over 5 years, the houndsfield-units diminished and the reading was "Disappearance of metastatic lesion, in favor of a liver cyst". This "cyst" was punctured by the radiologist. Prof. Dr. Günther: No tumor cells could be seen. The patient is now alive, 6.7 years after the primary operation.

(7) M. F., female, 48 years

1975 tumor nephrectomy following hypernephroma (left kidney). July 1979 finding of liver-metastases using liver-scan, computer tomogram, angiography and ultrasound. Unspecific immunotherapy started in March 1980 with 5 mg KLH given intradermally, followed by 4×10^6 allogenic radiated tumor cells. Finding of bone destruction in humerus-head R in form of a bone metastasis. After removal of metastasis, replacement of humerus-head with a prosthesis. Computer tomogram August 1980 shows no visible liver metastasis. Single kidney R shows tumor-like changes in some areas. A following angiography shows none of the earlier liver metastases, but definite pathological findings in the kidney R. Metastases in this patient are stable at this time. She is well-nourished and in only slightly reduced general condition.

Immunological Data

The immunoglobulin readings showed nothing below the norm in any of these patients. The effects of unspecific immunization with candida antigen was well demonstrated in the rising of the anti-body titer, as well as in the IgM and IgG-range.

T- and B-Lymphocytes

While the underpopulation of B-cells during immunization stayed stable, there was a T-cell depression during KLH-boosting, so far an unknown phenomena.

Lymphocyte Stimulation

Because of great breadth of dispersion, the results and the few patients, it is impossible to statistically rate the unspecific lymphocyte stimulation. On the whole, patients showed a tendency to lower transformation-rates, which could be improved by intradermal injection of candida or KLH, while changes in single mitogenes showed significant differences.

DNCB-Test

Five of 7 patients showed a negative DNCB-test, while after starting immunotherapy, 3 patients changed to a positive DNCB-test.

Discussion

Tumor nephrectomy is justified in patients who suffer not merely from bone-metastases, these are patients who also suffer from hypernephroma metastases and whose prognosis is unfavorable, where therapy on the whole is useless.

Tumor nephrectomy in these patients lengthened their lives by approximately 6 months [4]. After *Middleton* [7], 35 nephrectomy patients with multiple metastases lived not longer than patients without nephrectomy at the same tumor-stage.

Contrary to publications in Anglo-Saxon literature, radical tumor nephrectomy and extirpation of solitary metastases, was no advantage to 14 patients and their life-expectancy. The 3-year survival time was given at 5%. More so, hopes started concentrating on the so-called immunotherapy. What little information we have thus far, experienced through patients, keep us guessing that there are specific immunoreactions. In an immunostaging program, *Cole et al.* [1] proved the presence of a big antigen-pool, if a specific toxicity was present (per ex.: large tumor or metastases). They used allogenous cells in their microcytotoxicity-assay.

Lymphocytes of healthy people did not show these qualities.

Cummings et al. [2] reached similar results. A good target-cell lysis was found especially with the autologous system.

Kjaer [5] correlated the immuno reactivity of leucocytes in tumor patients, against an autologous tumor, allogenic tumor-cell pool and foetal kidney-tissue. He found a definite connection between better life-expectancy and reactions in leucocytic-migration-inhibition test.

Patients with a low leucocytic-migration-inhibition index (no metastases), had the same life-expectancy rate (result after 3-years' observation) as patients with primary metastases and a good migration-index. With the results of these tests, *Kjaer* managed to identify a "high-risk group". These important test results stimulated the use of various immunotherapies.

Morales [9] spoke about 4 transitory remissions of metastases in 10 patients suffering from metastasizing hypernephroma. BCG was given intradermally, weekly or monthly. Eight months was the longest observation-period.

Skinner et al. [11], after immunization of rams with hypernephroma cells, ribonucleinic acid was taken from their leucocytes and given as an tumor-specific immunotherapy to patients suffering from hypernephroma. Three of the so-treated patients with metastasizing tumors lived longer than 12 or 18.5 months with metastases staying stable.

Montie [8], 10 patients with metastasizing hypernephroma were given a polypeptide, which is able to restore certain immunodefects. Five of these patients showed temporary stabilization of metastases. Time of observation used – only 6 months.

Tykkae et al. [13] produced a prospective randomized 5-year study about active immunotherapy, involving autologous tumor cells, given in combination with immunostimulators BCG or candida-antigen. The amazing result of 23.6%, compared with 4.3% of the patients living past the 5-year limit, shows that immunotherapy is the best alternate way of treating hypernephroma. A control study undertaken by *Neidhard et al.* [10] proved altogether the results of *Tykkae*. Autologous and autogenic vaccinations were used in 30 patients. 7% showed complete regression of metastases, 37% showed stable metastases ("stable disease"). Half of the patients showed favorable reactions to immunotherapy, while the disease kept progressing in the other half.

Knowledge Gathered from These Tests so Far:

(1) Extirpation of primary tumor is a first valuable step towards immunotherapy.

(2) Metastases should be in their lowest, possible numbers.

(3) Immunization with autologous cells might be better than immunization with allogenic cells.

(4) Sole immunization with immunostimulators (per ex.: BCG, KLH) seems without effect.

(5) Mechanism in vivo is still to be cleared, especially T-cell depression as "second-set" phenomenon.

(6) Post-operative immunization seems more favorable than immunization at a later date.

A combination of cell-innoculation with candida-antigens shows the immunological reaction of the body towards fungal-antigens, by means of hemagglutination titer. It is easy to assume, that if a body builds up antibodies against fungal-antigens, it will have an immuno reaction towards tumor-antigenic cells, but this is only a hypothesis.

Reason for additional stimulation with immunostimulators, is the "alarm effect" candida-antigen has on the T-cell system, which through activation with a strong antigen, is supposed to be able to recognize weak tumor-antigens ("wake-up effect"). All this must be considered as hypothesis, as well.

Because of heterogeneity and varied therapy, these 7 case histories leave us with no general conclusion. It is important to know though, that, during treatment there was no tumor enhancement. Because of differences in dates, technical measurements and interpretations, it is not possible to come up with a definite therapy plan.

However, it does seem justified, that different centers having the necessary equipment, facilities and manpower, should undertake further controlled studies to definitely determine if immunotherapy can be a "big step forward" in treating metastasizing hypernephroma.

Summary

In 1976, an immunotherapy study in 7 patients with primary metastatic disease of renal-cell carcinoma was started. Follow-up is now at 102 months. Four patients received allogenic, three patients autologous tumor cell vaccination. Three patients died 22 months after beginning the therapy. Four patients were alive; 2 with stable disease. In 2 other patients, 57 months after the initial immunotherapeutic regimen, no lung metastases could further be seen. A metastatic liver lesion disappeared in one patient. This is a follow-up study of datas published in 1981 (Akt. Urol. *12:* 161–165, 1981) of patients of the Urological Clinic, University of Mainz (FRG).

Zusammenfassung

1976 wurde eine Immuntherapie an 7 Patienten mit primär metastasierendem Nierenzellkarzinom begonnen. Der Follow-up beträgt bis jetzt 102 Monate. 4 Patienten wurden allogen, 3 Patienten mit autologem Tumorzellmaterial vakziniert. 3 Patienten starben 22 Monate nach Beginn der Therapie. 4 Patienten sind am Leben und zwar mit stationärer Erkrankung. Bei 2 anderen Patienten konnten 57 Monate nach der ersten immuntherapeutischen Behandlung keine Lungenmetastasen mehr nachgewiesen werden, bei 1 Patienten bildeten sich Lebermetastasen zurück.

References

1 Cole, A. T.; Avis, I.; Fried, F. A.; Avis, F.: Cell mediated immunity in renal cell carcinoma – preliminary report. J. Urol. *115:* 234–238 (1976).
2 Cummings, K. B.; Peter, I. B.; Kaufman, J. J.: Cell mediated immunity to tumor antigens in patients with renal cell carcinoma. J. Urol. *110:* 31–35 (1973).
3 Jacobi, G. H.; Schneider, H. M.; Marberger, M.: Primär nicht-urologische Erscheinungsformen des hypernephroiden Nierenkarzinoms: Die primäre Metastase. Urologe A *17:* 64–72 (1978).
4 Johnson, D. E.; Kaessler, K. F.; Samuels, M. L.: Is nephrectomy justified in patients with metastatic renal carcinoma? J. Urol. *114:* 27–29 (1975).
5 Kjaer, M.: Prognostic value of tumor directed cell mediated hypersensitivity detected by means of the mocyte migration technique in patients with renal carcinoma. Eur. J. Cancer *12:* 889–898 (1976).
6 Klippel, K. F.; Altwein, J. E.: Palliative Therapiemöglichkeiten beim metastasierten Hypernephrom. Dt. med. Wschr. *104:* 28–31 (1979).
7 Middleton, R. G.: Surgery for metastatic renal cell carcinoma. J. Urol. *97:* 973–977 (1967).

8   Montie, J. E.; Bukowski, R. M.; Deodkar, S. D.; Hewlett, J. E.; Stewart, B. H.; Straffon, R. A.: Immunotherapy of disseminated renal cell carcinoma with transfer factor. J. Urol. *117:* 553–556 (1977).
9   Morales, A.; Eidinger, D.: Bacillus Calmette-Guerin in the treatment of adenocarcinoma of the kidney. J. Urol. *115:* 377–380 (1976).
10 Neidhart, J. A.; Murphy, S. G.; Hennick, L. A.; Wise, H. A.: Active specific immunotherapy of stage IV in renal carcinoma with aggregated tumor antigen adjuvant. Cancer *46:* 1128–1132 (1980).
11 Skinner, D. G.; de Kernion, J. B.; Brower, P. A.; Ramming, K. P.; Pilch, Y. H.: Advanced renal cell carcinoma: treatment with xenogenic immune ribonucleic acid and appropriate surgical resection. J. Urol. *115:* 246–250 (1976).
12 Talley, R. W.: Chemotherapy of adenocarcinoma of the kidney. Cancer *32:* 1062–1065 (1973).
13 Tykkae, H.; Kjelt, L.; Oravisto, K. J.; Turunen, M.; Tallberg, Th.: Disappearance of lung metastases during immunotherapy in five patients suffering from renal cell carcinoma. Scand. J. resp. Dis. (Suppl.) *89:* 123–128 (1974).
14 Tykkae, H.; Oravisto, K. J.; Lehtonen, T.; Sarna, S.; Tallberg, T.: Active specific immunotherapy of advanced renal cell carcinoma. Eur. Urol. *4:* 250–258 (1978).

Prof. Dr. med. K. F. Klippel, Chefarzt der Urologischen Klinik, Allg. Krankenhaus, Siemensplatz 4, D-3100 Celle (FRG)

AIDS – Clinical Picture and Therapy – Report on Two Patients

N. Vetter, F. Muhar

IInd Internal Pulmonary Department of the Pulmological Center, Vienna, Austria

The clinical picture of acquired immunodeficiency (AIDS) is characterized by the occurrence of unusual opportunistic infections and malignant symptoms (especially Kaposi sarcomas) in previously healthy adults. The disorder is associated with an immunodeficiency expressed in a change in the T-helper/T-suppressor cell-ratio in favor of the T-suppressor cells. In particular, the disorder appears in homosexuals, drug addicts and hemophilia patients and, after an uncharacteristic prodromal stage with fever, weight-loss, diarrhoea, and lymph-node swellings, displays an unfavorable course with the appearance of opportunistic infections. The cause of the disorder has not been clarified [6].

A causal therapy of the disorder is, consequently, not possible. In the clinical reports, we examine the clinical picture and the diagnosis of opportunistic infections. We report on the course of the disorder and our therapy. In addition to a specific therapy of opportunistic infections therapy of the cellular immunological defect was attempted in earlier studies. We report on an additional therapy with NeyThymun®, a thymus extract.

Method

NeyThymun® f and k are cytoplasmic thymus extracts of foetal thymus and juvenile thymus. The polypeptides contained therein (alpha-1 thymosin, thymopoietin and serum-thymus factor have, so far, been biochemically analyzed) should induce a lymphocyte differentiation in vitro and in vivo [3]. The substances were gradually applied in increasing concentrations (dilutions I, II, III corresponding to 10^{-12} g/ml, 10^{-9} g/ml, 10^{-6} g/ml) with a triple application of a dry substance (10^{-3} g/ml) in the first treatment-cycle and a continuous application three times per week of dry substances in the second treatment-cycle [7].

In the course of the treatment, serum parameters were repeatedly checked (blood picture with differential blood count, transaminases, BUN, creatinine, electrophoresis and immunoglobulins, virus serology, T-cells and subsets). The determination of T-cells, T-helper and T-suppressor/cytotoxic cells, was carried out with the aid of monoclonal antibodies (leu 4, leu 3, leu 2). The number of mononuclear cells (per cent) binding the leu 4, leu 3 and leu 2 was determined by immunofluorescence analyses by FACS.

Case Reports

L. A., 28 years

Anamnesis: The patient is homosexual and spent five years in New York. Two years ago, the patient reported non-specific symptoms with vitality deficiency, rapid fatigue and alternating lymph-node swellings. High temperatures began to occur one year ago, recently being manifested as septic fever attacks. In January 1983, profuse diarrhoea occurred for one week and in April 1983 there were difficulties in swallowing and vomiting; at this time, a thrush oesophagitis was diagnosed. At the end of June 1983, the patient reported unproductive coughing.

Clinical symptoms and findings on admission: The patient was in a reduced general state of health and nutritional condition (177 cm, 50 kg B. W.) and showed rales over the lungs and enlarged inguinal lymph nodes. The thorax X-ray displayed infiltration-free pulmonary fields and raised diaphragms and atelectases were demonstrated basally on the right. The pulmonary function showed a restrictive ventilation disorder with a vital capacity (VC) of 3.5 l (78%); in 1982, the VC had still been 6.1 l. There was no increase in the endobronchial resistances. The BSR was greatly increased with 110/135, erythrocytes 4.07 mill., leukocytes 4600, 5% stabs, 67% polymorphs, 6% monocytes and 22% lymphocytes in the differential blood count. Bronchoscopy revealed no abnormalities in the bronchial system and a bronchial lavage showed 100% alveolar macrophages. No "inflammatory cells" could be demonstrated, nor was it possible to show Pneumocystis carinii in the lavage liquid. The transbronchial pulmonary biopsy showed the histological picture of a Pneumocystis carinii pneumonia. It was possible to demonstrate the infectious agent in the cytological impression preparation. Pseudomonas aeruginosa and acinetobacter were demonstrated in the bronchial secretion; the blood culture was sterile.

Course: Despite broad-spectrum antibiotic therapy and high-dosage therapy with trimethoprim/sulfamethoxazole (720 mg trimethoprim, 3600 mg sulfamethoxazole daily), there was initially a deterioration in the clinical symptoms and in the X-ray picture: the patient displayed a tachypnoea of 50/min, an initially normal blood oxygen pressure showed a decline (pO_2: 64 torrs). Indistinct infiltrates appeared in the middle lobe and left lower lobe. After a brief corticosteroid therapy (50 mg prednisolone daily), there was a clear improvement in the clinical symptoms and in the changes apparent in the X-ray. After 4 weeks, the vital capacity had risen to 5.4 l (120%) and the respiratory rate and blood oxygen pressure were normal. Broncho-alveolar lavage showed 92% alveolar macrophages, 6% lymphocytes and 2% granulocytes, no Pneumocystis carinii could be demonstrated; it was decided not to take another pulmonary biopsy. The difficulty in swallowing of the thrush oesophagitis was still present, as before.

E. M., 28 years

Anamnesis: His grandfather, an uncle and the latter's family developed tuberculosis in 1960. The patient is addicted to heroin since 1971; a hepatitis B occurred in 1971 and a pulmonary tuberculosis in the left upper lobe in 1979. At the end of 1982, the patient reported nocturnal perspiration as the only symptom; at the end of November 1982, changes in the left upper lobe were diagnosed within the framework of heroin-withdrawal therapy; a sputum culture at this time showed the growth of Mycobacterium tuberculosis. A tuberculostatic therapy was started at the end of December 1982. The patient is homosexual and reports numerous sexual contacts with different partners.

Clinical symptoms and findings: When admitted at the end of January 1983, the patient was free from clinical symptoms. The BSR was normal with 4/15, all other serum parameters checked, including the liver function samples, were normal. The gamma globulin fraction was high (21.3 g/l), the IgG with 2660 mg/100 ml very high and the IgA and IgM in the normal range. The X-ray showed a cavity in the apical upper lobe segment; there were dense foci and also fairly small, indistinct, pale spotty shadows.

Course and therapy: Through a chemotherapy with INH, ethambutol and rimactan, a negativity of the spatially limited upper-lobe tuberculosis was achieved with a clear regression in the changes shown in the X-ray. In April, a loss in weight of 4 kg occurred, and a patchy, blue lesion at the knee appeared. Histological examination showed a moderate, but dense, proliferation of round, lymphoid and spindle cells around the vessels of the upper vascular plexus; the vessels in these areas were not characterized by a normal development. Erythrocytes were squashed in between the spindle cells, there were likewise small, vascular lumina of an incipient irregular shape around a hair follicle and around sweat gland glomerules and, to an increased perivascular extent, roundish and spindly cell elements. It was considered highly probable that it was a question of a Kaposi's sarcoma in the early maculate stage.

Results

The therapy and the course of the T-helper/T-suppressor cell-ratio are shown in figure 1 and 2.

The clinical course was favorable in both cases: after an initial deterioration in the clinical symptoms and in the X-ray findings of the Pneumocystis carinii pneumonia, one patient still suffered from difficulty in swallowing after the disappearance of the pulmonary symptoms. The difficulties in swallowing were less severe than on admission and there was a clear improvement in the oesophagus X-ray. A localized retinitis, which was the expression of a cytomegalo-virus infection, regressed in the course of the disorder. Subfebrile temperatures were still present as before. There was a noteworthy blood picture with this patient: we attributed the development of an anemia to the long therapy with a folic-acid antagonist (decline in the erythrocyte count to 3 mill. during the therapy with no change in the leukocyte counts or in the differential blood picture). The

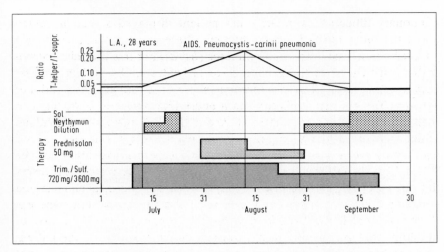

Fig. 1. Therapy and course of the ratio in a patient with Pneumocystis carinii pneumonia.

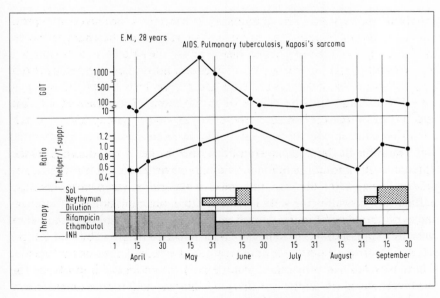

Fig. 2. Therapy, course of the ratio and of the GOT in a patient with pulmonary tuberculosis.

pulmonary tuberculosis in the other patient displayed a regular course under the tuberculostatic chemotherapy. There was a rapid regression in an intercurrent rise in the transaminases (fig. 2), but this caused a considerable impairment to the general state of health of the patient. A new manifestation of a Kaposi's sarcoma was not observed. The ratio of T-helper/T-suppressor cells showed a phase-like course in both cases: it rose with one patient from 0.03 to 0.25 and then fell again to 0.02 With an initial value of 0.5, the other patient showed a rise to 1.2, a drop to 0.4 and another rise to a maximum value of 0.9. The percentage counts of the T-lymphocytes displayed only slight fluctuations in the course of the disorder. In the phase of the rise in the ratio, there was a decline in the percentage counts of the T-suppressor cells with one patient, whereas there was a rise in the T-helper cells with the other patient. The course of the immunoglobulins displayed a constant IgG increase with one patient while with the other there was a constant IgA increase with unimportant fluctuations in the course of the disorder.

Discussion

With one patient, the pulmonary tuberculosis displayed a typical course with rapid sputum negativity and regression of the changes shown in the X-rays. A solitary lesion, histologically described as a "suspected Kaposi's sarcoma", was completely excised and no new lesion appeared within the course of 6 months. With the tuberculosis it was, however, a question of an apical tuberculosis of slight extent. In the case of the skin lesion, of a very early stage of the disorder with a patchy lesion (patch stage).

The other patient displayed a Pneumocystis carinii pneumonia which initially took a progressive course despite a trimethoprim/sulfamethoxazole therapy, was favorably influenced by a brief corticosteroid therapy and disappeared after 4 weeks of treatment. Four months after the occurrence of the Pneumocystis carinii pneumonia, only the symptoms of a thrush oesophagitis could still be clinically demonstrated, but subfebrile temperatures were still present as before. The early diagnosis by invasive methods (transbronchial pulmonary biopsy) was of crucial importance for the favorable course of the disorder. This gives a clear improvement in the prognosis of the disorder [5]. The importance of the broncho-alveolar lavage for the demonstration of Pneumocystis carinii has already been pointed out. With our patient, the demonstration of the infectious agent in the broncho-alveolar lavage was not successful.

In the 1st cycle of therapy with NeyThymun®, a rise in the T-helper/T-suppressor cell-ratio was seen in both patients; whereby with one patient a rise in the ratio (0.9 with an initial value of 0.5) was already demonstrated before the start of the therapy. After the application of the 2nd cycle, there again was a rise in the ratio of one patient while there was no change in that of the other, this continuing to be pathological. The constantly favorable clinical course did not correlate with the phase-like course of the ratio. Attention has been drawn to this phenomenon in the literature on the subject [2, 5].

Experimental studies on animals exposed to radiation showed a favorable course in an animal treated with NeyThymun® dil. st. III, in comparison with control animals with a less-marked decline in the lymphocyte count and a faster regeneration of B-cells [1]. The possibility of influencing the T-helper/T-suppressor cell-ratio was not studied. A therapy with thymosin and transfer factor [2] for an AIDS patient with a Mycobacterium avium intracellulare infection showed a rise in the T-helper/T-suppressor cell-ratio with a brief clinical improvement. Three months after the start of this therapy, the patient died from generalized cytomegalo-virus infections and infections with Mycobacterium avium cellulare.

Attempts have also previously been made to carry out an immunostimulation therapy with BCG, interleukin and interferon [4]; an adequate quantity of data has only been published on the interferon therapy. In a publication described as provisional, alpha-interferon was used for patients with Kaposi's sarcoma. Three of 13 patients displayed a complete remission (total disappearance of all symptoms over a period of one month, confirmed by biopsy). The average ratio of the T-helper/T-suppressor lymphocytes (OKT 4/OKT 8) rose from 0.63 to 0.96.

Summary

The course of the syndrome of acquired immunodeficiency cannot be influenced in long-term by any of the therapy concepts applied so far. We report on 2 patients in whom a therapy of opportunistic infections was successful. An early diagnosis by invasive methods is of paramount importance with Pneumocystis carinii pneumonia. A chemotherapy of a pulmonary tuberculosis proved effective with a rapid sputum negativity and a rapid regression of the changes demonstrated by X-rays. The favorable course so far with 2 patients is sufficient reason for us to maintain an additional therapy with NeyThymun®.

Zusammenfassung

Wir berichten über 2 Patienten mit dem Syndrom der erworbenen Immundefizienz (AIDS). Bei einem Patienten bestanden eine Soor-Ösophagitis und eine Pneumocystis-carinii-Pneumonie, bei dem anderen eine Lungentuberkulose und der Verdacht auf ein Kaposi-Sarkom im frühen Stadium. Wir berichten über den Verlauf und die Therapie der Erkrankungen und weisen auf die Bedeutung einer raschen Diagnose und Therapie opportunistischer Infektionen hin. Wir diskutieren über eine zusätzliche Therapie mit einem Thymusextrakt (NeyThymun®).

References

1. Buschmann, H.: Beeinflussung eines Strahlenschadens auf das Immunsystem durch Behandlung mit Revitorgan®-Präparaten aus fetalem Thymus (NeyThymun®) und Plazenta. Therapiewoche *33:* 191–196 (1983).
2. Elliot, J. L., et al.: The acquired immunodeficiency syndrome and mycobacterium avium-intracellular bacteremia in a patient with hemophilia. Ann. int. Med. *98:* 290–293 (1983).
3. Goldstein, A. L., et al.: Proc. Natn. Acad. Sci. USA *74:* 725 (1977).
4. Krown, S. E., et al.: Preliminary observations on the effect of recombinant leukocyte A interferon in homosexual men with Kaposi's sarcoma. New Engl. J. Med. *308:* 1071–1076 (1983).
5. Mildvan, D., et al.: Opportunistic infections and immunodeficiency in homosexual men. Ann. int. Med. *96* (Part I): 700–704 (1982).
6. Small, C. B., et al.: Community-acquired opportunistic infections and defective cellular immunity in heterosexual drug abusers and homosexual men. Am. J. Med. *74:* 433–441 (1983).
7. Stiefel, Th.: Thymushormone, Kassenarzt *22:* 27 (1982).

Dr. med. N. Vetter, II. Interne Lungenabteilung des Pulmologischen Zentrums der Stadt Wien, Sanatoriumstr. 2, A-1145 Wien (Österreich)

THE LIBRARY
UNIVERSITY OF CALIFORNIA
San Francisco
(415) 476-2335

THIS BOOK IS DUE ON THE LAST DATE STAMPED BELOW

Books not returned on time are subject to fines according to the Library Lending Code. A renewal may be made on certain materials. For details consult Lending Code.

INTERLIBRARY LOAN
DAYS AFTER RECEIPT
SEP 14 1989

RETURNED
OCT 11 1989